Love, work and knowledge are the well-springs of our life.
They should also govern it.

WILHELM REICH

CONTENTS

Chapters 1–4 in *Ether, God and Devil* and chapter 4 in *Cosmic Superimposition* have been newly translated from the original German manuscripts by Therese Pol. All other parts of these works were written in English by Reich. In this edition, minor editorial revisions have been made.

Mary Higgins, Trustee
Wilhelm Reich Infant Trust Fund

New York, 1973

ETHER, GOD AND DEVIL

What is the hardest thing of all?
That which seems the easiest
For your eyes to see,
That which lies before your eyes.

Goethe

CHAPTER I

THE WORKSHOP OF ORGONOMIC
FUNCTIONALISM

The cosmic orgone energy was discovered as a result of the consistent application of the functional technique of thinking. It was these methodic, rigidly controlled thought processes that led from one fact to another, weaving—across a span of about twenty-five years—seemingly disparate facts into a unified picture of the function of nature; a picture which is submitted to the verdict of the world as the still unfinished doctrinal framework of *Orgonomy*. Hence it is necessary to describe the "functional technique of thinking."

It is useful not only to allow the serious student of the natural sciences to see the result of research but also to initiate him into the secrets of the workshop in which the end product, after much toil and effort, is shaped. I consider it an error in scientific communication that, most of the time, merely the polished and flawless results of natural research are displayed, as in an art show. An exhibit of the finished product alone has many drawbacks and dangers for both its creator and its users. The creator of the product will be only too ready to demonstrate perfection and flawlessness while concealing gaps, uncertainties and discordant contradictions of his insight into nature. He thus belittles the meaning of the real process of natural research. The user of the product will not appreciate the rigorous demands made on the natural scientist when the latter has to reveal and describe the secrets of nature *in a practical way.* He will never learn to think for himself and to cope by himself. Very few drivers have an accurate idea of the sum of human efforts, of the complicated thought processes and operations needed for manufacturing an automobile. Our world would be better off if the beneficiaries of work knew more about the *process* of work and the experience of the workers, if they did not pluck so thoughtlessly the fruits of labor performed by others.

In the case of orgonomy, a look into a corner of the workshop is particularly pertinent. The greatest difficulty in understanding the orgone theory lies in the fact that the discovery of the orgone has solved too many problems at once, and problems that were too vast: the biological foundation of emotional illnesses, biogenesis and, with it, the cancer biopathy, the ether, the cosmic longing of the human animal, a new kind of physical energy, etc. There was always too much going on in the workshop; too many facts, new causal connections, corrections of dated and inaccurate viewpoints, connections with various branches of specialized research in the natural sciences. Hence, I often had to defend myself against the criticism that I had overstepped scientific limits, that I had undertaken "too much at one time." I did not undertake too much at a time, and I did not overreach myself scientifically. No one has felt this charge of "too much" more painfully than I have. I did not set out to trace the facts; the facts and interrelations flowed toward me in superabundance. I had trouble treating them with due attention and putting them in good order. Many, many facts of great significance were lost that way; others remained uncomprehended. But the essential and basic facts about the discovery of cosmic orgone energy strike me as sufficiently secure and systematized for others to continue building the structure I could not complete. The multitude of new facts and interrelations, particularly the relationship of the human animal to his universe, can be explained by a very simple analogy.

Did Columbus discover New York City or Chicago, the fisheries in Maine, the plantations in the South, the vast waterworks, or the natural resources on America's West Coast? He discovered none of this, built none of it, did not work out any details. He merely discovered a stretch of seashore that up to then was unknown to Europeans. The discovery of this coastal stretch on the Atlantic Ocean was the key to everything that over several centuries became "North America." Columbus's achievement consisted not of building America but of surmounting seemingly immovable prejudices and hardships, preparing for his voyage, carrying it out, and landing on alien, dangerous shores.

The discovery of cosmic energy occurred in a similar fashion. In reality, I have made only one single discovery: *the function of orgastic plasma pulsation.* It represents the coastal stretch from which all else developed. It was far more difficult to overcome human prejudice in dealing with the biophysical basis of emotions, which are man's deepest concern, than to make the relatively simple observation about bions or to cite the equally simple and self-evident fact that the cancer biopathy rests on the general shrinking and decomposition of the living organism.

"What is the hardest thing of all? / That which seems the easiest / For your eyes to see, / That which lies before your eyes," as Goethe put it.

What has always astounded me is not that the orgone exists and functions, but that for over twenty millennia it was so thoroughly overlooked or argued away whenever a few life-asserting scholars sighted and described it. In one respect, the discovery of the orgone differs from the discovery of America: orgone energy functions in all human beings and before all eyes. America first had to be found.

An essential and comprehensive part of my activities in

the workshop lay in learning to understand why people in general, and natural scientists in particular, recoil from so basic a phenomenon as the orgastic pulsation. Another part of my work, which brought down on me much dirt, dust, and plain garbage, consisted of feeling, experiencing, understanding, and overcoming the bitter hatred, among friends and foes alike, that formed a roadblock everywhere to my orgasm research. I believe that biogenesis, the ether question, the life function and "human nature" would long ago have been conquered by many scientific workers if these basic questions of natural science had not had but *one* access: the orgastic plasma pulsation.

When I succeeded in concentrating on this single problem for three decades, mastering it and orienting myself within its fundamental natural function, in spite of all obstacles and personal attacks, I began to realize that I had transcended the conceptual framework of the existing human character structure and, with it, our civilization during the past five thousand years. Without wanting to, I found myself *outside* its limits. Hence I had to expect that I would not be understood even if I produced the simplest and most easily verifiable facts and interconnections. I found myself in a new, different realm of thought, which I first had to investigate before I could go on. This orientation in the new functional realm of thought, in contrast to the mechanistic-mystical realm of patriarchal civilization, took about fourteen years, roughly from 1932 to the writing of this work, 1946 and 1947.

My writings have often been criticized for being far too compressed, forcing the reader to make a strenuous effort at concentration. It has been said that people prefer to enjoy an important book in the same way they enjoy beautiful scenery while cruising at leisure in a comfortable car. People do not want to race toward a specific goal in a straight line at lightning speed.

I admit that I might have presented *The Function of the Orgasm* in a thousand instead of three hundred pages, and the orgone therapy of the cancer biopathy in five hundred instead of one hundred pages. I further admit that I never troubled to familiarize my readers completely with the conceptual and investigative methods on which the results of orgonomy are based. No doubt this has caused much damage. I claim extenuating circumstances insofar as I opened up several scientific fields over the decades, which I first had to set down in a condensed, systematic form in order to keep up with the development of my research. I know that I have built no more than the scaffold and foundation of my structure, that windows, doors, and important interior features are missing in many places, and that it does not offer a comfortable abode.

I ask to be excused because of the pioneer nature of this basically different research. I had to gather my scientific treasures rapidly, wherever and however I found them; this happened during the brief intervals between six changes of domicile forced upon me partly by "peaceful" circumstances but partly by extremely violent social upheavals. Furthermore, I constantly had to start from scratch in earning a living: first in Germany (1930), then in Copenhagen (1933), in Sweden and Norway (1934, twice in the same year), and in the United States (1939). In retrospect, I ask myself how I succeeded in accomplishing anything essential at all. For almost two decades I lived and worked "on the run," so to speak. All this precluded a congenial and secure atmosphere, without which it is impossible to give congenial, extensive descriptions of discoveries. I must reject another criticism, namely, that I unnecessarily provoked the public by the word "orgasm" in the title of a book. There is no reason whatever for being ashamed of this function. Those who are squeamish about it need not read further. The rest of us cannot allow others to dictate the limits of scientific research.

When I began this book, I planned to make up for what I had denied to myself and others for so long in terms of breadth and more graphic presentation. I hope I will now be spared the criticism that I have taken my research "too seriously" by giving it "too much" space.

Since everything in nature is interconnected in one way or another, the subject of "orgonomic functionalism" is practically inexhaustible. It was essentially the humanistic and scientific achievements of the nineteenth and early twentieth centuries that merged with my interests and studies of the natural sciences to form the living body of work that eventually took useful and applicable shape as "orgonomic functionalism." Although the functional technique of thinking will be described here systematically for the first time, it was nevertheless applied by many scholars more or less consciously before it definitely overcame, in the form of orgonomy, the hitherto rigid limits of natural research. I would like to mention the names of those to whom I am primarily indebted: Coster, Dostoevsky, Lange, Nietzsche, Morgan, Darwin, Engels, Semon, Bergson, Freud, Malinowski, among others. When I said earlier that I found myself in a "new realm of thought," this does not mean that orgonomic functionalism was "ready" and merely waiting for me, or that I could simply appropriate Bergson's or Engels's conceptual technique and apply it smoothly to the area of my problem. The formation of this thought technique was in itself a task I had to accomplish in practical activity as a physician and scientist struggling against the mechanistic and mystical interpretations of living matter. Thus I have not developed a "new philosophy" that adjacent to, or in conjunction with other philosophies, tried to bring the processes of life closer to human comprehension, as some of my friends believe. *No, there is no philosophy involved at all.* Rather, we are dealing with a tool of thought that we must learn to apply before investigating the substance of life. Orgonomic functionalism is not some luxury article to be worn or taken off at one's discretion. It consolidates the conceptual laws and functions of perception that must be mastered if we are to allow children and adolescents to grow up as life-affirmative human beings in this world, if we want to bring the human animal into harmony again with his natural constitution and the nature surrounding him. One can oppose such a goal on philosophical or religious grounds. One can declare, "purely philosophically," that a "unity of nature and culture" is impossible or harmful or unethical or unimportant. But no one can claim any longer that the splitting up of the human animal into a cultural and a private being, into a "representative of higher values" and an "orgonotic energy system," does not, in the truest sense of the word, undermine his health, does not harm his intelligence, does not destroy his joy of living, does not stifle his initiative, does not plunge his society time and again into chaos. The protection of life demands functional thinking (in contrast to mechanistics and mysticism) as a guideline in this world, just as traffic safety demands good brakes and flawlessly working signal lights.

I would like to confess to the most rigid scientific ordering of freedom here. Neither philosophy nor ethics but the protection of social functioning will determine whether a child of four can experience his first genital excitations with or without anxiety. A physician, educator, or social administrator can have only *one* opinion (not five) about the sadistic or pornographic fantasies a boy or girl develops during puberty under the pressure of moralism. It is not a question of philosophical possibilities but of social and personal necessities to prevent by all possible means the deaths of thousands of women from cancer of the uterus because they were raised to practice abstinence, because thousands of cancer researchers do not want to acknowledge this fact or will not speak up for fear of ostracism. It is a murderous philosophy that still favors the suppression of natural life functions in infants and adolescents.

If we trace the origins and wide ramifications of public opinion, especially with respect to the personal life of the human masses, we find time and again the ancient, classic "philosophies" about life, the state, absolute values, the universal spirit. They are all accepted uncritically in an era that has degenerated into chaos because of these "harmless" philosophies, an era in which the human animal has lost his orientation and self-confidence and senselessly gambles away his life. Thus, we are not concerned about philosophies but about practical tools crucial to the reshaping of human life. What is at stake is the choice between good and bad tools in rebuilding and reorganizing human society.

A tool alone cannot do this work. Man must create the tools for mastering nature. Hence it is the *human character structure* that determines how the tool will be made and what purpose it will serve.

The armored, mechanistically rigid person thinks mechanistically, produces mechanistic tools, and forms a mechanistic conception of nature.

The armored person who feels his orgonotic body excitations in spite of his biological rigidity, but does not understand them, is mystic man. He is interested not in "material" but in "spiritual" things. He forms a mystical, supernatural idea about nature.

Both the mechanist and the mystic stand inside the limits and conceptual laws of a civilization which is ruled by a contradictory and murderous mixture of machines and gods. This civilization forms the mechanistic-mystical structures of men, and the mechanistic-mystical character structures keep reproducing a mechanistic-mystical civilization. Both mechanists and mystics find themselves inside the framework of human structure in a civilization conditioned by mechanistics and mysticism. They cannot grasp the basic problems of this civilization because their thinking and philosophy correspond exactly to the condition they project and continue to reproduce. In order to realize the power of mysticism, one has only to think of the murderous conflict between Hindus and Muslims at the time India was divided. To comprehend what mechanistic civilization means, think of the "age of the atom bomb."

Orgonomic functionalism stands outside the framework of mechanistic-mystical civilization. It did not rise from the need to "bury" this civilization; hence, it is not a priori revolutionary. Orgonomic functionalism represents the way of thinking of the individual who is unarmored and therefore in contact with nature inside and outside himself. *The living human animal acts like any other animal, i.e., functionally; armored man acts mechanistically and mystically. Orgonomic functionalism is the vital expression of the unarmored human animal, his tool for comprehending nature.* This method of thinking and working becomes a dynamically progressive force of social development only by observing, criticizing, and changing mechanistic-mystical civilization from the standpoint of the natural laws of life, and not from the narrow perspective of state, church, economy, culture, etc.

Since, within the intellectual framework of mechanistic-mystical character structure, life itself has been misunderstood, abused, feared, and often persecuted, it is evident that orgonomic functionalism is outside the social realm of mechanistic civilization. Wherever it finds itself inside this realm, it must step out of it in order to function. And "functioning" means nothing but investigating, understanding, and protecting life as a force of nature. At its inception, orgone biophysics possessed the important insight that the functioning of living matter is simple, that the essence of life is the vital functioning itself, and that it has no transcendental "purpose" or "meaning." The search for the purposeful meaning of life stems from the armoring of the human organism, which blots out the living function and replaces it with rigid formulas of life. Unarmored life does not look for a meaning or purpose for its existence, for the simple reason that it functions spontaneously, meaningfully, and purposefully, without the command "Thou shalt."

The interrelations between conceptual methods, character structures, and social limitations are simple and logical. They explain why, so far, all men who understood and battled for life in one form or another consistently found themselves *frustrated outsiders*—outside the conceptual laws that have governed human society for thousands of years—and why they so often suffered and perished. And where they seemed to penetrate, it can be consistently shown that the armored exponents of mechanistic-mystical civilization time and again deprived their doctrine's life-affirmative element of its specific characteristics and embodied it into the existing conceptual framework by diluting or "correcting" it. This will be discussed at length elsewhere. Here it suffices to prove that functional thinking is outside the framework of our civilization because life itself is outside it, because it is not investigated but misunderstood and feared.

CHAPTER II

THE TWO BASIC PILLARS OF HUMAN THOUGHT: "GOD" AND "ETHER"

Whoever looks back on the development of human society a thousand or five thousand years from now will very likely find the crucial turning point of human orientation in our own time, the twentieth century. Removed from its emotional turmoil, and from a higher vantage point, he will perceive the large outlines that shaped the errors of the human animal. He may also discern the first faint beginnings pointing toward his own time. The standpoint from which the human animal will examine and judge his history and his present will be the science of life. What has dictated the life forms of the human animal today and in the past six millennia will be a matter of historical criticism. What finds itself today outside civilization will be the judge of the past, if my functional technique of thinking is correct. This is not a prophecy but an inevitable logical conclusion. For if it is correct that the mechanistic and the mystical life philosophies have, each in its own way, attacked the living element in the human animal, if it is correct furthermore that mechanistics and mysticism derived their methods of thinking from the *negation* of life, it must be equally correct that their collapse will be brought about by the discovery of the life process. Even today it is clear that both mechanistics and metaphysics have cruelly failed as tools of human existence.

Now it is a rule of development that false conceptual systems continue to exist until their bankruptcy leads them to develop new conceptual systems that then take over to guide human destiny. Bankrupt thought systems first drown the life of the human animal in blood and tears before any life-assertive thought is stimulated to ensure life itself. If an organism is dying, the life function struggles to the last breath in mighty convulsions—the death agony—against ultimate stillness. By the same token, society fights against strangulation from false conceptual systems by creating new ones that in the light of existing ideas may seem "revolutionary" or "radically new." Upon careful examination they turn out to be desperate attempts to revive very old ideas that could not prevail at the time or were deprived of their vitality by the sluggish-thinking human masses. The energy observed in the agonal struggle of a dying animal is not "basically new" or "alien" or energy from another source; it is the same innate life energy that drove the organism to look for food and to enjoy life. Likewise, the modes of thought that were erroneously described as radical or revolutionary and led to a new social order in times of crisis were not newly introduced or concocted; they can be traced to the very beginnings of human organization. It is not hard to establish the fact that they are even older than those thought systems which they tried to overcome time and again, often in vain. This is true for both the mechanistic and mystical worlds of ideas. We find the accent on life thousands of years ago—in the ancient thought systems of the great Asiatic religions such as Hinduism, certainly in early Christianity, and in the beginnings of the natural sciences in antiquity.

The position of life, of the biological, is therefore not "new" and does not have to be introduced. It is the *oldest* position in human thought; one is even tempted to say that it is the most conservative. This raises the logical question of why it remained so powerless and was displaced by other thought systems that time and again drove humanity into disaster. Today, and certainly to our observer five thousand years from now, it must seem very strange that, in spite of their cruelty and futility, the life-negating thought systems could persevere and torture mankind. How this could happen is indeed a question that requires an answer. Can it be given?

At present, I am merely trying to outline the wide range covered by this book. When I confronted the task of formulating the principles of orgonomy and their underlying thought technique, I faced a dilemma:

Orgonomy is the science of the functional laws of cosmic orgone energy. There were two ways of organizing the material: the one was academic, or "detached"; the other was human, or "involved." Involved in what? Mainly in the objective accuracy of scientific observations, facts and interconnections. Certain functions of nature, hitherto unknown, had to be described and defined. In the process of this important work, I was time and again disturbed by one specific question: *Why did man, through thousands of years, wherever he built scientific, philosophic or religious systems, go astray with such persistence and with such catastrophic consequences?* Scientific skepticism is necessary and justified. As natural scientists we are professional nonbelievers because we know man's vast capacity for error, the unreliability of his impressions and the enormous area of erroneous judgments.

And yet the question is justified and necessary: Is human erring necessary? Is it rational? Is all error rationally explainable and necessary?

If we examine the sources of human error, we find that they fall into several groups:

Gaps in the knowledge of nature form a wide sector of human erring. Medical errors prior to the knowledge of anatomy and infectious diseases were necessary errors. But we must ask if the mortal threat to the first investigators of animal anatomy was a necessary error too.

The belief that the earth was fixed in space was a necessary error, rooted in the ignorance of natural laws. But was it an equally necessary error to burn Giordano Bruno at the stake and to incarcerate Galileo?

Reason tells us that we cannot find rational, comprehensible grounds for the burning of Bruno and for similar massacres—no immediate grounds, that is. The answer that "This is how it always was" is no answer at all but merely an expression of sluggish thinking.

Though the human animal errs in its concept of nature, it nevertheless erects a conceptual structure that, while wrong, is inherently consistent. The inner logic of erroneous thought systems is comparable to the inner consistency of a paranoid delusion. Both the conceptual and the delusional systems are even related to some part of reality. But in both the thinking departs at a certain point from objective reality and develops an "inner logic of errors" of its own.

We understand that human thinking can penetrate only to a given limit at a given time. What we fail to understand is why the human intellect does not stop at this point and say: "this is the present limit of my understanding. Let us wait until new vistas open up." This would be rational, comprehensible, purposeful thinking. What amazes us is the sudden turn from the rational beginning to the irrational illusion. Irrationality and illusion are revealed by the intolerance and cruelty with which they are expressed. We observe that human thought systems show tolerance as long as they adhere to reality. The more the thought process is removed from reality, the more intolerance and cruelty are needed to guarantee its continued existence. This cannot be explained by saying, "People happen to be like that." This is no insight at all; however, the stubbornness of such a pretext betrays a secret meaning. Let us try to find it.

New systems of thought are intended to overcome the errors of the old systems. As we well know, the former spring from the contradictions of erroneous thinking, which has become divorced from reality. Since it derives from this realm of thought, the new system carries some of the old sources of error into its new structure. The new thought system grows in the logical, although incorrectly arranged, old one. The pioneering intellectual spirit is socially tied to his own time. To this is added a purely psychological factor, namely, that the pioneer does not want to discard altogether the erroneous conceptual system, with its comfortable, familiar features. He would like to be understood in his own time, without having to stand entirely outside it. Hence he neglects to use his critical faculties, takes over erroneous concepts, or disguises his innovations by using dated, sterile words. The new system vacillates, too insecure to find its own ground. But the old system which is under attack enjoys public applause and contemporary esteem; it is backed by organization and power.

To illustrate our point let us examine several of the great human errors in natural science.

The great revolutionary astronomer Copernicus developed his conceptual system from the critique of the Ptolemaic system. Yet he took over Ptolemy's concept of the static center of the universe by substituting the sun for the earth. He took over the idea of the "perfect" world system, which, in his time, was presented in the form of a perfect circle and uniform motion. With the idea of "perfection," the idea of the "divine" was introduced into the new system. We will see the continuing effect of this error for centuries. It was composed of three erroneous concepts: the "perfect," the "divine" and the "static."

If we look for realistic proof of these three elements in his system of thought, we cannot find a single supporting fact. Yet this grandiose error must originally have had a function.

Kepler developed his own system of thought from the framework of the Copernican system. He corrected the Copernican perfect circle of the earth around the sun, stretching it into an ellipse and thus breaking the idea of perfection. While Kepler demonstrated no process in nature that caused the earth's path to become elliptical, he nevertheless formulated his three laws of planetary motion, which are valid to this day. As will be shown later, orgonometry follows his harmonic law, confirms it and makes it comprehensible in terms of energy. Kepler also did away with perfect, uniform motion by proving that, in its course around the sun, the earth's radius vector moves over equal areas in equal times. Hence the earth moves fastest where it is nearest to the sun —at its perihelion—and slowest where it is most distant from the sun—at the aphelion.

As we shall see later, Kepler even found a clue to cosmic orgone energy when he asked himself which force was responsible for the attraction of the earth to the sun. He already knew about the energy field; he knew that the sun also rotates, and the field with it; but his sun is fixed in space. It does not move on its own; it is the final frame of reference for the earth's motion.

Although the error of the sun's fixed position was subsequently corrected, it still persists in practice. Today's calculations of planetary motion still begin with the concept of the static sun. Otherwise Kepler's functional law, which describes an ellipse with the sun at its focus, could not be applied. After all, the ellipse, like the circle, is a *closed* geometric figure. The sun in motion as the center of the planetary system would logically preclude the idea of an elliptical path of the planets; it would *open* the path of the planets.

If we look for the core of these vast errors, we encounter the *static* element time and again. It continuously permeates all scientific concepts. Even Kant, who corrected so many gross errors of his time, who raised the function of human knowledge itself to the goal of natural research, allowed the static element, in the form of the ABSOLUTE metaphysical moral principle, to sneak in. It seems as if human thinking cannot—or will not—shed the static notion. There is invariably something resting and motionless in the abundance of things in motion. It reemerges in the contemporary idea of "cosmic dust." It appears to be absolute, independent of any functional process, as if it were a part of eternity. What does all this mean? The inaccuracy of such concepts is self-evident, including their proof, which still remains to be furnished.

Anyone who attempts to trace the error that permeates all thinking should first ask himself if his own conceptual perspective is foolproof. He should ask if there are any guarantees against basic errors in thinking, and what they consist of. The result of this needed self-criticism is contained in this book, in which the basic conceptual principles of orgonomy will be examined. They can be summarized roughly as follows:

1. There are basic principles of a thought technique that are objectively valid and controllable.

2. There is a logical development in finding facts of nature that will prove the applied conceptual technique to be right or wrong. A theory is correct only if it leads to the discovery of new basic facts (not only to details of known facts).

3. The factual foundation of a correct conceptual technique relates to the evolved theory as the base of a pyramid relates to the apex. The larger the scope of logically interwoven facts in relation to the theory, the more reliable is the latter. The smaller the factual basis in relation to the theory, the faster the theory will collapse. Conversely, we should reject as unscientific those thought techniques in which a broad theory relates to the facts

as the pyramid base relates to the apex, or those
techniques that have no factual foundation at all.

4. The scientific observer must know his own perspective lest he make incorrect statements. He must
know in which functional realm of nature he and
his objective are placed.

5. The scientist will increase his errors in proportion
to the neglect of his own system of sensory perceptions and awareness. He must know how he himself
functions when he perceives and thinks.

If we accept these criteria of correct thinking, we have
also gained access to one of the most important sources of
human error: *the ignorance of the scientist or thinker with
regard to his own conceptual system and his sensory perceptions.* In other words, every previous thought technique
of the human animal was in danger of attributing to nature
certain qualities of human structure that cannot be found in
a given object of nature; the same is true for the other danger, namely, to evade functions in the human structure that
are unknown or discredited, even though they, too, are
found in nature.

The complete failure to recognize an existing, basic,
cosmic energy must be attributed to the effects of the second
danger.

If I say "failure to recognize an existing, basic, cosmic
energy," I mean its natural scientific comprehension, not
merely theoretical formulations about it; I mean its *factual*
proof, and not intellectual constructs without factual foundation. This distinction is important if I am to lend credibility to my assertion that the human animal has known about
the existence of a primordial cosmic energy since the beginning of recorded history. I would like to elaborate on this
point because it is a part of determining my own position in
the investigation of nature.

I noted in my introduction that my research and conceptual technique were outside the ways of thinking of the past five thousand or six thousand years of civilization. WHERE "outside"? This was not easily established. Until after I had discovered orgone energy in the atmosphere and had comprehended its general cosmic nature, I could not know that I was inside the realm of basic, cosmic functions when I discovered and described the function of the orgasm.

But at the start of my work this positive orientation of my scientific perspective was lacking. I could tell only where I was *not*, not where I was. To work without knowing his positive standpoint is an intolerable burden to the scientist. Hence it is understandable that I, along with many of my scientific contemporaries, was clutching at straws in the chaos, under the illusion that I could in that way make myself secure. Thus I committed several gross and dangerous errors in my thinking. The fact that I was finally able to free myself from them should be attributed to my discovery of the much-maligned function of the orgasm rather than to my intellectual superiority at that time. These conceptual errors were characterized by a chaotic abundance of contradictory prejudices and mistaken judgments. I shall try to group them synoptically, according to the sequence in which I committed them.

Let us organize the realms of human thought according to their increasing scope. The following schematic drawing will illustrate my point:

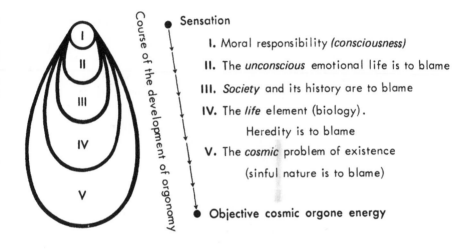

Diagram of the realms of human thought and their objective *interdependence*

The social life about the time of the First World War
(1914–1918), in spite of Marx and Nietzsche, was governed by the *idea of guilt,* by *absolute morality.*

The "moral existence" of the human animal was subject
to *conscious moral responsibility.* Everything evil that happened was blamed on the "evil will" of man. The damage
caused by Schopenhauer to human thought affected intellectual circles just as much as did the "original sin" of the
Church. Man as a *conscious* being was held fully responsible
for morality and justice, and as a citizen for his thoughts and
acts. To this day he has not freed himself of this erroneous
thinking. There was no logical question as to where consciousness and will originate. Consciousness, will and responsibility were indestructible metaphysical conditions, existing
from time immemorial, absolutely and forever. Nietzsche's

magnificent critique of morality had no social impact. It was not a part of the *Zeitgeist,* such as will and guilt. But it paved the way toward an important step—the step into the realm of *unconscious psychic life,* investigated and comprehended by Freud. One should emphasize that while organized religion made use of guilt in the moral consciousness, guilt was not actually anchored in man's consciousness. I shall return a little later to this matter.

Depth psychology put an end to the absolute nature of consciousness, along with *conscious* guilt, by proving that consciousness reflected the *unconscious* psychic life. But it fed such notions as the absolute, the eternal and the guilt concept from the realm of moral consciousness back into the unconscious. Now human guilt was no longer rooted in conscious malevolence and immorality but in the *instinctual drives of the unconscious.* The child appeared as a "polymorphous perverse" human being, as a "wild animal" that had to be tamed and "adjusted to culture." This was the enormous error in the *second* realm of knowledge.

Today, the "guilt of the malevolent unconscious" in man dominates the thinking of a large public. Because of the mysteriously stubborn wrong-headed thinking of the human animal, we must suspect that we will have to drag along this error for centuries to come. By definition, such errors tend to become fixed, to widen their sphere of influence as far as possible and to anchor themselves in dogma, without proof. For the next logical question is: *where do the malevolent unconscious drives originate?* In what functioning realm of nature is the rationality of thinking anchored? From which function of nature stems the famous human mind according to which the "malevolent" animal in the infant must become adjusted to "high culture"? In psychoanalysis, both intellect and unconscious drive appear as gigantic, static, eternal configurations. There is neither a whither nor a yon. This error of depth psychology, which was clinically refuted by sex-economic investigations, is supported because only a few depth psychologists have any useful knowledge of natural science in general and of the biosociological interdependence of emotional problems in particular. Feeding on the emotional misery of the human animal, a large number of psychologists, except for those who remained fixed in the old realm of consciousness, barricaded themselves behind the absolute, static processes of the malevolent unconscious and defended every attack on this bastion by all available means, except scientific ones.

Of course, I do not deny the existence of malevolent unconscious emotions in the human animal, and I have often said so at length. But in my concept, man is part of the rest of nature. Hence his malevolence is placed into another functional system which has an origin, a reason for being where it is, and an end, like all natural functions.

The question of whether man is fundamentally a "good-natured" or a "malevolent" animal is irrelevant. We are not concerned with moral theology. We want to know what place man holds in the totality of nature, with his "good" and "evil" drives. This is our position in the investigation.

If we try to give a more exact description of nature, we encounter two other large systems of thought—*sociology* and *biology* (III and IV in our diagram). If we place ourselves temporarily outside prevailing opinions, the answer seems simple, even commonplace: Somewhere, sometime, life began to differentiate from the rest of nature. What we call human society today developed after millions of years from the wide realm of life. It is a specifically differentiated part of the living realm, just as that realm is a specifically differentiated part of nature. This conclusion is correct even if we cannot say anything as yet about the *how* of these two differentiations and their internal functions. The reverse conclusion, namely, that nature is part of the living realm or even a principle of society would be absurd. As an individual, man is subject to the laws of life and to his social circumstances. In the primeval past, an infernal mixture of false ideas disturbed this simple relationship between man and nature. To this day, the human animal has not been able to free itself, and in the future this large third error will presumably victimize just as many human lives as it did in the past millennia.

All genuine natural science stands outside the given social framework; it judges the essence of human existence in the vast context of nature. The given social framework does not harmonize with man's objective position in nature but reflects the erroneous concepts that, in the course of time, the human animal has formed about its position in nature.

We find a great part of this enormous error in the inability of the masses of people even to think about their position in nature; in their tendency to follow blindly the heresies taught by individuals and, beyond that, to persecute and torture anyone who tries to clarify this error. In the twentieth century, the masses have not transcended the state of dull, merely vegetating existence into which they have fallen since those tremendous errors took hold. The problem is the hate-ridden rejection of all fundamental knowledge of nature.

Let us review the crude details of these enormous errors. Standing outside the conceptual framework of this civilization, we are not obliged to turn right or left to see if we would harm certain "interests" or offend personal sensibilities. Our perspective is too far removed to permit any offense to our personal integrity from any side. Neither are we interested in whether people believe us or not, since this cannot possibly shake our logical thought processes.

One of the most glaring errors is that, without knowledge of himself, *the human animal has drawn conclusions about itself and applied them to the essence of nature.* This is true not only for the so-called prescientific era of antiquity and the Middle Ages but also, very acutely, for the present. Now this is an error of basic thinking. Since man is a part of nature, and not the other way around, he can only draw conclusions from nature, and never the reverse. Even where we study the perceptual and conceptual apparatus of the human animal itself in order to learn how we perceive the world around us, we must investigate the natural functions in man. In other words, we must derive sensory perception itself from natural, physical processes and must not examine them outside the processes of nature.

Aside from the unthinking masses of human animals who succumb to social suffering, we find that the large conceptual systems, including their errors, were accomplished only by individual human animals. They did not think in a vacuum. Their questions about the existence of man and nature were dependent upon their biological and social existence. Nevertheless they did confront these functioning realms and thought about them. The ability to think in itself must have astounded them. "Cogito—ergo sum!" The history of natural philosophy is the history of this astonishment about the ability to think, to perceive and to judge, right up to Kant and to the dull classes in logic at secondary schools and universities.

Surprised by his own ability to think, man was trapped by the erroneous conclusions about himself and applied them to nature. We understand this surprise and the subsequent false conclusions. But we do not understand his stubbornness and the cruelty of his wrong-headed insistence on it. We understand the origin of the idea that the earth is the center of the universe and that Mount Olympus is peopled by all manner of gods. But we do not understand the murderous hatred toward anyone trying to correct these false ideas.

We understand how the idea of *Homo sapiens* developed. But so far no one has plausibly explained why, in the Middle Ages, the study of the human organism was forbidden. This

is all the more incomprehensible since the human animal presented himself on every occasion as the living incarnation of
his ideas. Rationally, man, first and foremost, should have
studied his own organism. The anthropomorphic view of
nature is far older than the mechanistic one. It is also far
older than knowledge about man. Today, in the twentieth
century, this knowledge is scarcely one hundred twenty years
old.

We pose questions in order to define our standpoint and
its surroundings. We have not even tried to furnish answers.
We only want to know the nature of the area in which we
build our scientific home. It is not our objective to explain
this environment. We only want to know as much as possible
about its qualities. The more we question, the more we are
surprised. We are amazed at the abundance of conceptual
errors committed in the course of a mere few thousand years.
We do not gloat; we feel very humble: *Whence springs the
enormous tendency to err?* Attitudes of stoical superiority
and "detached" philosophy will not do. Unless we succeed
in tracking down the compulsion toward error, it is pointless
to add one more error to the many others. Human animals
talk and write far too much in a vacuum anyway. Those concerned with the function of knowledge itself must exercise
the most rigorous self-control. No other approach can be
taken seriously. Let us investigate further.

It is permissible to claim that psychology is man's oldest
method for orienting himself in his environment. It is certainly no coincidence that until quite recently psychology
was taught in conjunction with philosophy. The study of nature was intimately connected with the study of emotional
life. The machine age did not develop any natural philosophy, but it introduced the mechanistic viewpoint into psychology and natural philosophy.

When I say "mechanistic," I mean a still undefined mixture of various concepts grouped around matter and its mo-

tion. Until the discovery of radium, roughly forty to fifty years ago, matter appeared to be static, visible, palpable, unalterable, ruled by the law of the "conservation of matter," moved by a "force," absolute, eternal in the form of atoms and "cosmic dust." The *absolute* and the *static* were taken over even by such dynamically oriented psychological schools as Freud's in the form of given unconscious ideas. With Jung, the unconscious emotional life became the static "archaic unconscious" and the static "collective unconscious." Together with the static viewpoint, many branches of psychology, even after they broke with philosophy, took over the problem of *guilt*. This drove them into a blind alley, from which there was no exit. For to give up the static, absolute standpoint of the emotional and sensory apparatus is tantamount to giving up psychology as a science of ultimate natural functions. The next logical consideration invariably is this: the emotional elements cannot have existed from time immemorial; they must have developed. With this consideration, both the material and the static viewpoint collapse. Development is, by definition, a dynamic process. Hence, there is no longer any firm resting point; all has slipped into flux. For someone psychologically oriented only toward the static and absolute concepts, this means losing his psychic bearing. If the archaic unconscious and the biologically absolute Oedipus concept are no longer the responsible "culprits," who then remains to take over the guilt (= original sin)? WHO THEN IS RESPONSIBLE FOR THE TREMENDOUS MISERY?

So far we have been in the second domain, that of the unconscious emotional life. We are now entering the *third* domain, that of the *social human community*. Here too we encounter the absolute and the static, the "scapegoat." We are still investigating, without trying to understand anything.

As before in the psychic and moral realms, we again encounter the idea of original sin: the absolute depravity of the human animal is also to blame in the social domain. If man were not so lecherous, carnal and sinful, there would be paradise. It is intriguing to speculate why a human system of thought that has lasted through the millennia never penetrated far enough to question where lust, carnality and sinfulness came from. If they are absolute and eternal concepts, if the Son of God had to be crucified to free mankind from its grave guilt, divine creation cannot possess the degree of perfection attributed to it. The "divine" creature called "man," who is also deeply "sinful," is a flagrant contradiction in terms. No doubt the scholastic-clerical viewpoint describes some kind of reality. But even this reality is static and eternal, in terms of guilt and sin, in the concept of eternity and in the idea of God. We are also on safe ground if we assume that the idea of guilt established the power of the Church, and not the other way around. Therefore, what does all this nonsense of "eternal guilt" mean?

Religion, with its metaphysical error of absolute guilt, dominated the broadest, the *cosmic* (V.), realm. From there, it infiltrated the subordinate realms of biological, social and moral existence with its errors of the absolute and of inherited guilt. Humanity, split up into millions of factions, groups, nations and states, lacerated itself with mutual accusations. "The Greeks are to blame," the Romans said, and "The Romans are to blame," the Greeks said. So they warred against one another. "The ancient Jewish priests are to blame," the early Christians shouted. "The Christians have preached the wrong Messiah," the Jews shouted and crucified the harmless Jesus. "The Muslims and Turks and Huns are guilty," the crusaders screamed. "The witches and the heretics are to blame," the later Christians howled for centuries, murdering, hanging, torturing and burning heretics. It remains to investigate the sources from which the Jesus legend derives its grandeur, emotional power and perseverance.

Let us continue to stay outside this St. Vitus dance. The longer we look around, the crazier it seems. Hundreds of minor patriarchs, self-proclaimed kings and princes, accused one another of this or that sin and made war, scorched the land, brought famine and epidemics to the populations. Later, this became known as "history." And the historians did not doubt the rationality of this history.

Gradually the common people appeared on the scene. "The Queen is to blame," the people's representatives shouted, and beheaded the Queen. Howling, the populace danced around the guillotine. From the ranks of the people arose Napoleon. "The Austrians, the Prussians, the Russians are to blame," it was now said. "Napoleon is to blame," came the reply. "The machines are to blame!", the weavers screamed, and "The lumpenproletariat is to blame," sounded back. "The Monarchy is to blame, long live the Constitution!" the burghers shouted. "The middle classes and the Constitution are to blame; wipe them out; long live the Dictatorship of the Proletariat," the proletarian dictators shout, and "The Russians are to blame," is hurled back. "Germany is to blame," the Japanese and the Italians shouted in 1915. "England is to blame," the fathers of the proletarians shouted in 1939. And "Germany is to blame," the self-same fathers shouted in 1942. "Italy, Germany and Japan are to blame," it was said in 1940.

It is only by keeping strictly outside this inferno that one can be amazed that the human animal continued to shriek "Guilty!" without doubting its own sanity, without even once asking about the origin of this guilt. Such mass psychoses have an origin and a function. Only human beings who are forced to hide something catastrophic are capable of erring so consistently and punishing so relentlessly any attempt at clarifying such errors.

The answer rested in the biology of the human animal. But he had already closed the access to this domain, too. He had barely started to think in terms of biology when he blocked all progress by falling into the vast error of "hereditary predisposition and hereditary degeneracy." Now it was not only the Jews, Japanese, Christians, Huns, Russians, capitalists, Negroes who were "to blame for all this calamity" but "genetic traits" were responsible. Children who were ruined by sick parents were condemned for having a "hereditary taint." Drinkers who took to the bottle to escape their social plight were regarded as having an "inherited tendency to drink." Women who sold their bodies because they were hungry or found no gratification within the Catholic concept of original sin were thought to be hereditary degenerates. Neurotic persons who could not earn a living were "genetically inferior." Mental patients, the victims of an education that had already crushed what is alive in the infant, were "hereditary degenerates." The black men living close to nature in Africa were "sinful" and "in need of salvation"; but the white slaver was the epitome of normality.

Our psychiatric knowledge about the functions of the human animal tells us that it fought against painful self-awareness when it accused the millions of victims of its brutality and grotesque errors of having "bad heredity." "Bad heredity is at fault," said the biologists (they are still looking for the genes of "criminal" sexual intercourse during puberty!), pathologists, psychiatrists, legislators, exponents of social medicine—against all the evidence, against daily experience, against the massive numbers of the sick and the dead caused by this error.

The human animal felt deep inside itself a degeneracy, a deviation from all nature. But, incapable of penetrating to the heart of the matter, it put the blame on the victims of its own degeneracy. What prevented the human animal from penetrating to its own core?

When the biological armoring was discovered, all the hostile camps united: psychoanalysts and communists, communists and fascists, biologists and pathologists who blamed hereditary factors; briefly, all those who owed their social existence to the great erring united to destroy the germinating knowledge about the general biological degeneracy. If these groups had used only a fraction of their energy for fighting pornography on the newsstands of all cities in the world, for fighting political hacks and hangmen, instead of battling against truths, all ideas of guilt would long have vanished from the world. This was the error in the fourth, or biological, realm.

If our assumption is correct, namely, that the great errors in human thought systems are connected with the concepts of the static-absolute and with guilt, we face two principal tasks at the start of a new scientific orientation:

1. We must investigate why the human animal, contrary to all sensory experiences in nature, invariably clings to the static absolute, i.e., to the immobile, to guilt. This is the task for psychology.

2. We must determine if the "absolute" corresponds to any reality in objective nature.

Let us return to the social realm of thought. Here we find an error which, to my mind, surpasses for sheer grotesqueness and perversity all other errors committed in the entire history of human development.

The critique of the absolute moral and psychological standpoint in the realm of social thinking was accomplished essentially by Karl Marx. Without denying it as such, he suspended the absolute and eternal character of the moral and psychic existence of the human animal by reducing it to the social conditions of life. Scientifically, this was a correct and great achievement. Our third, the social, realm of life happens to be smaller than the biological, but wider than the psychological, conscious or moral domain. The psychic and moral existence is put into the social context and constantly draws the content of its ideas from this function, and by no means the other way around. The science of the social realm of life is "sociology" or "history." It cannot work with quantities, except for statistical studies, but it rests on a conceptual system, the sociological one. In this conceptual system the psychic-moral element was relativized, i.e., reduced to interpersonal relations. Since the founder of this doctrine worked at a time of prospering capitalism, it is understandable that he emphasized the capitalistic social structure. Yet he was also wise enough to relativize this capitalistic structure, namely, to seek its origin in earlier social organizations (feudal, etc.). But the social thought system did not penetrate into the fourth system, the biological, of which it is a partial function. That the founder of this social thought system guessed at the interdependence between the social and the biological element can be proved by his assertion that "the social process was a process of nature." Marx never held the individual capitalist responsible for social abuses, even though he openly exposed the brutality with which the proletariat was then treated (child labor, injustice, etc.). Now something grotesque happened:

The human animal adopted this thought system and reintroduced the absolute and static elements. Now the "capitalist" was, and is, to blame. The human animal banishes, murders, hangs and tortures thousands of capitalists while retaining, in camouflaged form, the very system it set out to destroy; it organizes constant terror against the individual and his freedom of thought, enlarges and absolutizes the idea of the state, which it intended to overthrow, and now puts the blame on capitalism in foreign countries. "The capitalist is to blame"—this has become the absolute, the static, the vast conceptual error. The absolute state is everything, man is nothing. Where the Church had needed centuries to subvert its original great idea, red fascism needed only a few years to ruin a great doctrine of human emancipation and turn it into its opposite, the most vicious mass deception in the history of mankind.

We are still asking questions. We are deliberately acting like a stranger in a faraway city. We want to find out why things are organized as they are, and not differently. The longtime resident of a city would not dream of asking such questions. To him, everything is routine and self-evident. Within the framework of his physical and psychic existence everything seems to be understood. There are no basic problems. The traffic jams in the streets are no problem to the native New Yorker. "That's the way it is," he will tell you. "That's New York for you," he will say. He does try to "minimize the trouble" by prohibiting parking in certain streets or by ordering one-way streets. New York "happens" to be a city of teeming millions; hence the traffic jam during rush hours is a feature of daily life.

Within their own realms, all conceptual systems are logical and correct, similar to delusional systems. That they are incorrect is obvious only to the out-of-town visitor who asks silly questions. For instance: Why have eight million people crowded together on an island so small that they were forced to build skyscrapers in order to save space? The New Yorker is astonished by such questions and thinks they are stupid. The outsider says: America is vast. Thousands of acres of land are uninhabited. Why did eight million people insist on settling in Manhattan? The New Yorker can "explain" it. New York happens to be "a metropolis," or "That's the way it is . . ." or "It's great to live among so many people . . ." But why? asks the stranger. Hundreds of thousands of children never see a tree or a field. The air is humid and polluted. All shops are overcrowded so that the purchaser has to wait a long time before being served. The apartments are small and poorly furnished, and expensive to boot. The same money would guarantee a better life in the vast areas of the country. "New York is the capital of the world," is the reply.

If one looks at human existence from the standpoint of its living quality, and not from the religious, industrial, governmental or cultural aspect, one asks simple questions that seem stupid, naïve or even crazy to the local church member, factory owner, statesman or president of this or that cultural organization.

About thirty years ago, when orgonomic functionalism began to ask the first naïve questions about human life, no one guessed that the issue of "what is life?" was being raised. The questions were simple and logical, and the answers were sharp and offensive to the world of the static and the absolute. Let us compile some of these naïve questions:

Why have all social programs of the political parties failed? Why create ever new programs only to see them fail again? Why are men unable to carry out their old and sound programs? One program is as well-intentioned as the other: the Christian love for one's fellow man; the idea of liberty, equality and fraternity; the American Constitution; the Constitution of 1848; Lenin's social democracy, etc., etc. The aspirations everywhere are the same, and so are the ideals.

If all men desire peace, then why is there always another murderous war, against the will and vital interests of the global population?

If there is no personal God, how can one explain the enormous power of all personalized religions? There must be a reason.

If nature has established sexual maturation during puberty, why is love forbidden at that age?

Why is it so hard for truth to assert itself against lies and defamation? Why is it not the other way around, that lies have to assert themselves against the truth?

In the United States, the doctrine of civil rights forms the basis of the Constitution. Why does the American Civil Liberties Union have to fight against the violation of civil rights, instead of political reactionaries having to defend themselves against civil rights?

Why are children so cruelly treated? Why are they swathed so tightly that they cannot move? Why are infants placed on their bellies so that they have trouble keeping their heads away from the pillow? Where does the general hatred of the child come from?

Why does man hate every new, correct thought? Surely his life would be better, and not worse, if he thought correctly. Does man think at all? Or is correct thinking a special talent?

How is it possible that millions of industrious people can be oppressed by a handful of rulers?

Why does the average person evade serious questions that go to the heart of the matter?

Why do people always discuss unessential, and never essential, matters in the United Nations? It is obvious that important matters and simple answers are avoided. Why?

Why is there a vote for some functionary in every corner of the world? Why is there no vote for peace or war, an issue that concerns the lives of millions?

Only fools or sages ask such simple questions. The answers are well known:

Political programs are needed because man is a "political animal."

Political programs fail because the politicians of the *other* party are so corrupt.

Human sinfulness, the nobility, the capitalist, the Bolshevik and the Jew are to blame for the unhappiness.

Children are tortured because they have to learn perseverance in the hard struggle for existence and because their "character must be hardened."

Adolescents must not enjoy love because they are too *immature* for it, or because they cannot enter into a marriage, or because they still have a lot to learn, or because "such things aren't done," or because it would hurt the development of their morality, or because "the family is sacred."

The persecution of truth has always existed, and always will. Hence there must be an ethic.

The trampling down of human rights is an evil the Civil Liberties Union was founded to fight.

People do not think, or think incorrectly, because they "just" vegetate.

To conduct plebiscites about war or peace is not "customary." Such proposals would not get a majority of votes.

In diplomatic circles it is "not customary" to ask important questions directly. To be diplomatic "happens" to be the essence of diplomacy.

People hate new ideas because they "happen to think" sluggishly or because they "happen" to be tradition-bound. That is how people "happen to be" . . .

The reader who has honestly thought about human life will now better understand why the true scholar and artistic creator is always outside the familiar. Not because he wants it that way, but because *he must be outside if he is to accomplish anything,* if he wants to avoid falling into the trap of the large errors of thinking.

In the course of his errors, the human animal has not stood still. He has shown a development even in his mistakes. He has evolved from the divine being, "Homo divinus," to the knowing man, "Homo sapiens." When it was proved to him that he was a bundle of irrational drives, he transformed himself into the final product of development, "Homo normalis." With every step of this development, the scope and depth of his errors have increased. The idea of *Homo divinus* was far less widespread and powerful than that of *Homo sapiens*. But *Homo sapiens* did not enjoy the power and honors of *Homo normalis*.

Those who want to comprehend the great error of "Homo normalis" must reach far back, because no other conceptual error has ever erected such high barriers against being understood.

If we disregard the *self-evident* reasons for human error, there remains a residue that is incomprehensible and bizarre: the murderous hatred of everything new and true, and of the natural function of love. *To this day, the deep aversion of the average person to questions touching the core of his life is not understood.* The counterpart of this aversion is man's penchant toward superficiality.

One can easily have many friends as long as one stays within the framework of well-established thought patterns. The friends run off as soon as this framework is transcended. Only very few will go along. The affability and helpfulness of people also cease when their given framework of thought is transcended.

We must reject "explanations" that merely serve to give a name to the matter, such as "human stupidity," "tradition," "influence of the Church," or "politics," or "the will to power" and similar notions. Such verbalizations express precisely what has to be explained—superficiality and evasiveness.

There must be a barrier somewhere, as if it were forbidden to touch on certain matters. What we are looking for cannot be what is commonly known as "sex." For the average person talks about nothing else so much, makes about nothing else so many bad jokes, as he does about "sex." It cannot simply be sexuality that is taboo. The matter goes deeper and is of a fundamental nature.

It is quite easy to distinguish two basic, comprehensive thought systems in the human mind. One is of a meta-

physical or mystical nature and centers on the idea of a supernatural being that shapes all events. It is the idea of
"God," which characterizes all religious systems, no matter
how much they may differ in detail. The other system postulates a physical force penetrating and dominating everything
that exists. This system is centered on the idea of an "ether."

In ancient Asiatic philosophies, the ether takes on the
properties of a living being without actually becoming one
—prana and similar concepts.

In addition to the two broad conceptual systems of *God*
and *ether,* there is a third, which has no connection with any
discernible processes in nature; it is characterized, most
clearly in Christianity, as the *devil.* For the moment, the
conceptual realm of the devil must be excluded because it
stems from morbid human fantasy. "God" and "ether," on
the other hand, are thought systems springing from the rational attempt of the human animal to comprehend his origin. Even now we may suppose that God and ether refer to
physical realities, while the devil is a totally irrational
concept. We do not yet know which role is given to the irrational.

The thought systems of "God" and "ether" form, each by
itself, a logical construct; they are opposites. The idea of
God derives everywhere from the inner psychic sensations;
the idea of the ether derives from rational thought processes
explaining physical phenomena. God supposedly explains the
emotional and spiritual existence of man, while the ether explains his material, physical existence.

Presumably these two thought systems originated independently from each other and have also been preserved independently from each other. But both conceptions are many
thousands of years old. They form the nucleus of the two
great systems of *religion* and *science.*

Although they apply to different realms of life and operate independently, they do have very remarkable features in common. Here is a grouping of their common features:

GOD	ETHER
is universal	is universal
is the source of all existence	is the source of all existence
is "perfect"	is "perfect"
is omniscient	is the basis of consciousness
is eternal	is eternal
is static	is static
is the mover and creator of celestial bodies	is the origin and mover of celestial bodies
is the origin of all matter and energy	is the origin of all matter and energy
is impenetrable	"cannot be proved"

Now let us look at the differences between the two ideas:

GOD	ETHER
Sensory life	Energy processes in nature
Spiritual	Physical (though unprovable)
Realm of religion, the subjective element	Realm of natural science, the objective element
Mysticism	Mechanistics

In human thinking, God and ether have something else in common, namely, that scientific thinkers regard them both as nonexistent. Materialistic natural philosophy, along the lines of La Mettrie, Buechner, Marx, Lange, etc., denies the existence of God. Einstein's school of physics denies the existence of the ether. But so far it has not been possible to replace the idea of God nor that of the ether by a useful concept of the nature and origin of life.

We started with the large errors in human thinking. We are trying to understand *why* people err so gravely and why they cling to these errors with such tenacity. We might easily proclaim another new theory about the universe. But we would not have the slightest guarantee that we might not

add a new error to the old ones. The greater the popular suc-
cess of our theory, the greater the damage it would cause. It
is no longer a question of new theories, any more than it is a
question of new political programs. We are concerned exclusively with finding the source of stubborn human erring. The
control of our conceptual technique is more important than
any other task. It is no accident that with the discovery of
irrationality in the human character the foundation of all
physical interpretations of the universe grew shaky.

Genuine natural science has always tried to test the accuracy of its judgments. One of the greatest methodological
difficulties lies in the fact that although it has to describe
objective functions of nature no judgment is independent
from individual sensory perception, and sensory perception
belongs to the *subjective* sensory apparatus of the scientist.
He is supposed to be "objective," without ever being able to
free himself from the subjective viewpoint. This basic difficulty in all scientific work is so great that influential schools
have often split over the question of whether there really is
an objective realm that can be perceived by the senses
(empiro-criticism) or even whether there is any reality at all
that exists independently of our perceptions and feelings
(solipsism). This idea was challenged by empiricism, which
had no qualms in accepting the outside world for what it
seems to be (positivism, mechanistic materialism). Added to
empiricism was the powerful system of metaphysical spiritualism, which gained the widest dissemination of all thought
systems. It sees no problem concerning the accuracy of our
statements about nature. It proceeds from subjective feelings alone and draws uncritical conclusions from man to an
absolute spirit in his own image.

The uncertainties in judging one's own perceptions and
conclusions have always been so great that one often felt that
famous philosophical schools had run into a roadblock of
compulsive brooding (e.g., Husserl). I am merely presenting

a brief survey because it is not my job to examine the scientific value of the various schools of thought. Such studies have often been undertaken by scholars with better knowledge of philosophy. My task will be limited to searching for the common principle guiding the typical human erring.

The critical reader will rightly ask what entitles me to pose as the judge of human error. It might be said that only a higher being has this right. Men should remain modest and admit that man happens to be a cruelly erring creature, from time immemorial unto eternity, and that only God is omniscient.

I am not so immodest as to present myself as an omniscient judge of human error. But I do claim the right to remain *outside* philosophical controversies and to ask why men, when they do think, utter so much nonsense although so much that is true and tangible is all around them. As a biopsychiatrist, who, for three decades, has judged and treated people from all cultural and social strata, I have earned the right to introduce a new viewpoint and to test its qualifications for limiting the field of *unnecessary* human erring.

I do not believe that my critics are genuinely modest. While they warn me to be modest and to shun the arrogance of passing divine judgments, they are presumptuous enough, every day, to describe the characteristics of God to millions of people, down to the last detail, without ever having seen God. And when the natural philosopher warns me to watch the limits of my sensory perceptions, I must answer: You have described the ether and endowed it with certain properties without ever having seen it. You have built world systems, purely in your mind, without ever perceiving or understanding a single element of "empty space." You have done away with the ether, thereby committing a colossal error. You have replaced a real, pulsating, lively, functioning world with numbers. I, on the other hand, have discovered by logical thought processes a force of nature that you have consistently overlooked or denied, and therefore I am entitled to question why you were so much in error and what role the natural force I discovered plays in natural phenomena.

I know that I can err, just as you can. But I do my best to protect myself from error by asking myself conscientiously how I happened to discover cosmic orgone energy. Since this force of nature is universal, I am trying to find if it is in any way connected with your ideas of "God" and "ether." In contradistinction to metaphysicians, theists and relativists, I rely on direct observations and controllable processes in nature. I am not responsible for the wide scope of the natural force I discovered. I take every available precaution to eliminate as many errors as possible. I do not philosophize and I make no statements about nature unless I have observations and controlled experiments on which to base my statements. Above all else I take account, exact account, of the relations between my own perceptions and the natural processes that are independent of myself. By placing myself outside all previous thought systems, I gain a perspective that has a chance of finding the common principle of error in all thought systems. I act like a bystander during a brawl in a saloon. I take no sides, and I take no part in the brawl. I am on the sidelines and ask myself what causes the brawlers to beat one another up. I ask if this apparently pointless fight is necessary or if it could have been avoided.

I am not trying to construct a world system, although I possess far more numerous and far more basic facts about nature than any other trend in natural science. It is possible that I, too, will be forced to outline a total picture of nature for myself. But this total picture will be merely a conceptual framework and I will build it only:

1. if I have understood why previous thought systems erred so gravely and so typically;

2. if the "world picture" in my mind is produced spontaneously from a multitude of controlled facts;

3. if I have previously understood the consistency in the sequence of my findings over three decades.

Such a consistency in discovering unknown functions must in itself be an important function of nature. It obviously affects the relationship between the natural scientist and nature which he has studied, and of which he is a part.

I do not believe that these safeguards can be found elsewhere in natural research. They are indispensable because the object of my research is all-encompassing and vitally important. Furthermore, I am ready to correct any of my errors.

So much for my attitude toward my own errors. The understanding of an investigative process spanning several decades, which began with the study of (much-maligned) pleasure sensations and ended with the discovery of a universally existing, unknown cosmic energy, is in itself indispensable and provides a safeguard against basic errors.

To reassure some religious-minded souls, I would like to add the following:

I do not claim to have discovered God or ether. I merely claim to have discovered a useful, practicable fact of nature that reveals many characteristics previously attributed to God and ether. I do not know if "God" or "ether" exists. But I do know that cosmic orgone energy has properties that I studied without reference to God or ether; properties that were completely unknown to me when I started my research, and equally unknown to other natural scientists; properties that I discovered piece by piece through observation and experiment, properties that came to me as the logical sequence of thought processes. I can further affirm that in 1941, when I had a long talk with Albert Einstein about my new discovery, the possible connection between orgone energy and the concepts of God and ether was remote from my mind. How remote it was, and how unbiased my work, may be proved by the fact that during our whole conversation the "elimination of the ether" and its substitution by equations did not even occur to me, although I knew about them. So I did not set out to discover ether or God. The discovery of facts about nature that formed the basis of human concepts of an invisible God and an ether was therefore objective and unintentional. Only the decision to remain outside all traditional thought systems, never to yield to prejudices, to follow strictly my observations wherever they might lead, to control rigorously my own conceptual technique and pay no heed to any authoritarian scientific claim, no matter from where it sprang, was intentional and carried out scrupulously. I can add to this my earlier determination to disregard any threat to my existence from Church, state or political parties that might endanger my line of research. I experienced the power of unbending search for truth with just as much surprise as a pathological dreamer and scoundrel in Germany must have experienced the social results of his consistent lies.

The search for truth is closely connected with the natural organization of the human animal. Hence, we may conclude that the evasion of truth and the adherence to the surface of phenomena must also have a certain connection with the structure of the human animal. The function of natural research must be buried somehow if the tendency to evade obvious facts is that powerful.

With this thought, I found the key to the riddle of why man could err so consistently, so cruelly, for so long, and so much to his own disadvantage: needless human erring is a pathological quality of human character.

Here some self-examination is again in order. Am I seduced by the idea that I am incapable of conceptual error?

Am I stupid or vain enough to assert that some stroke of fate gave me the qualities it denied to others, namely, not to err?

Anyone who is familiar with my writings knows that I am neither vain nor stupid enough to proclaim such arrogance. He knows that ever since I called myself a natural scientist, I have belonged among those who, at the risk of their lives, fiercely struggle against any kind of so-called God-given facts and flawlessness. I consider the arrogant pretense of a Russian "proletarian" despot to omniscience and omnipotence just as sick and harmful as I do the Pope's arrogance in proclaiming himself as the all-knowing, infallible representative of God. I am doing this from my scientific viewpoint, which blames neither the German nor the Russian nor the Catholic Father of all peoples, but solely the character structure of the human animal. As I have often explained, the root of misery should not be sought in the intentions or cruelties or arrogance of individual persons but in the biology of the human animal. "Human animals alone are responsible," I have proclaimed time and again, contrary to all public opinion.

I know human erring from my own experience. I, too, have joined in shouting "Guilty! Guilty!" It is helpful to refer to the schema on page 22 in order to demonstrate this erring.

I began to err when I held religion alone responsible for human suffering. I did not know that the error of religion was a symptom, not the cause, of human biopathy. I persisted in my error when I held the personal interests of a social group—parents or educators—responsible for suppressing human love life. I did not know that the suppression of love life is no more than a mechanism, and by no means the final cause, let alone intention, of certain social circles.

When I was under the spell of the great socialist movement and worked for years, as a physician, among the under-privileged strata of the people, I fell into the gross error of thinking that "the capitalist was responsible for human plight." It took the brutal experience of the deteriorating Russian Revolution to free me from this error. They had killed the capitalists, but misery continued to grow; diplomatic intrigues, political maneuvering, spying and informing on others, all of which they had set out to eradicate, were more powerfully at work than ever. These experiences inflicted deep wounds.

For years, and in harmony with Freud's doctrine, I committed the error of thinking that the unconscious was "evil" and "responsible for all misery." It took a full decade of hard, clinical work among the emotionally ill to free me from this error. This earned me the bitter enmity of many psychiatric businessmen, who enriched themselves at the expense of human emotional misery.

Thus I committed the gravest errors of my time and even defended them with conviction. But I do claim for myself that I did not cling to them, as did so many of my co-workers and professional colleagues. I have remained mobile.

Whether I am now falling into a new error, I do not know. I assume that it is correct to trace human distress to the pathology of human structure, which in turn lies in its armoring, and to hold the armoring responsible for the orgastic impotence of the human animal, but all this may be a mere mechanism. The answer lies somewhere in that area of our existence which has been so heavily obscured by organized religion and put out of our reach. Hence, it probably lies in the relation of the human being to the cosmic energy that governs him.

But even if I continue to make errors, I try to find their sources. By and large this can be done. And while I am not at all prepared to go along with human viewpoints that have been proved wrong, I am eminently capable of accepting cor-

rections of my errors and assimilating them into my thinking. I have proved this. I no longer believe, as I once did,
that the over-all guilt rests in the "evil will" of the first domain, or in the "evil unconscious drives" of the second, or in
the "evil capitalist" of the third, or in the fixed "hereditary
traits" of the fourth, or in the "sin against the Holy Ghost"
of the fifth domain. There is no guilt at all, but merely an
uncomprehended catastrophe in the biosocial development
of the human animal. His biological armoring stands out as
the central mechanism of his failure, but not as its cause.
Aside from the known mechanism and the known consequences, the armoring, too, must have a comprehensible origin.

From now on, we shall seek the origin of the tendency to
err in the armoring of the human animal. This armoring is
the only known function in man that is characterized by immobility. It works against the mobility of living functions
and originated as an inhibiting mechanism. The immobility
that strikes us as the hallmark of all human errors—the
static, the absolute, the immovable, the eternal—might very
well be an expression of human armoring. We could accept
this conclusion only if the essential traits of human error
were identical with the essential traits of the armoring, well
known from clinical observations. We would thus have
gained part of a secure foundation from which to judge our
scientific perspective. We would clearly differentiate between
the life expressions of the human animal, i.e., its motility,
and its armoring and the resultant blocking.

The thought technique underlying this process presupposes that the human animal cannot think, postulate or do
anything that is not somehow rooted in his biopsychic structure. According to this viewpoint, the biopsychic apparatus
of man is the medium through which all inner and outer functions have to pass before they become thoughts or acts.
To put it differently: *man can think or do nothing, no*

*matter how incorrect from the viewpoint of his own life or
from the objective insight of nature, that does not somehow,
somewhere, contain a nucleus of objective truth;* in other
words, that would not be meaningful or rational in some respect or other. Consequently, even the grossest human errors, such as the belief in supernatural spirits or in an
absolute creator of all being, have a rational function and a
comprehensible meaning. Even the guilt question and the
absolute have a function corresponding to some reality.

It is surprising to find that the human animal has proclaimed its two greatest idols—God and ether—to be virtu-
ally unknowable. One might also add the idea of "mankind"
to these "unknowables." Man has erred nowhere so gravely,
so frequently and so consistently as in the conceptual realms
of "God," "ether" and "man," although nothing else has occupied and stirred him as much as these three ideas. He has
failed in all three realms and found no practical support in
them. Since we presuppose that the ideas of "God," "ether"
and "man" are rooted in a reality, it follows logically that
the armoring of the human animal must be responsible for
his thinking in a vacuum in all three areas. Evidently, he perceives reality as in a mirror, without ever touching it.

Let us dare a further assumption:

The conceptual worlds of "God" and "ether" show so
many similarities that they must have a common origin, regardless of the fact that God as an *esthetic quality* and ether
as a *physical quantity* so far have never met, and could not
meet within the framework of human thought. There is no
bridge between God and ether in the thinking of the human
animal, just as there is no bridge between the beauty of a
color and its corresponding frequency of oscillations (vibra-
tions) in the ether. It seems all the more peculiar and significant because God and ether have so many similarities in
the world of human imagination.

The physical reality underlying the concepts of "God"

and "ether" might be the primal cosmic energy, orgone energy. And the motives that so far have prevented the human animal from finding and describing God as well as ether practically might be the same as those that have prevented it from discovering cosmic orgone energy. Following this assumption, if the living, dynamic element in the human animal perceived cosmic orgone energy, metaphysically as "God" and physically as "ether," his armoring prevented him from putting his perceptions to scientific and practicable, technical use.

This is the wide framework of thought in which the theme of this book will be elaborated. The following functional diagram will show these assumed relations. Proof of their actual reality will take us far into the realm of nature.

Perception	Energy
Soul	Body
Spirit	Matter
Metaphysics	Materialism
Mysticism	Mechanistics, technology
Religion	Science
Quality	Quantity
Subjective	Objective

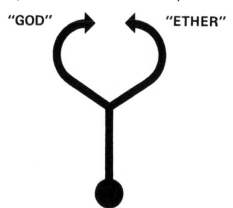

"GOD"   "ETHER"

COSMIC ORGONE ENERGY

Primordial energy
Universally existent
All-permeating
Origin of all energy (motion)
Origin of all matter
In the living being: biological energy
In the universe: origin of the galactic systems

Diagram of the functional relations between God, ether and orgone energy; framework of the conceptual technique of this book.

CHAPTER III

ORGAN SENSATION AS A TOOL OF
NATURAL RESEARCH

Pleasure, longing, anxiety, rage, sadness, roughly in that order, are the basic emotions of life. They are predicated on the completely free motility of the organism. Each of these emotions has its own particular quality. They all express a motile condition of the organism, which has a "significance" (psychologically: a "meaning") in relation to the self and the world at large. This significance is rational. It corresponds to real conditions and motile processes of the protoplasm. The primary emotions of life also have a rational function. The function of pleasure leads to the discharge of surplus cell energy. Anxiety is at the base of every reaction of rage. And in the realm of life, rage has the over-all function of conquering or eliminating life-threatening situations. Sadness expresses the loss of familiar contact, and longing expresses the desire for contact with another orgonotic system. We shall have to show later that it is the function of emotion that constitutes the goal of a drive, and not the reverse, as the metaphysicians postulate. Here, we merely intended to demonstrate that primary emotions are, and must be, rational if life is to function "meaningfully." This is proven by its existence.

I emphasize the rationality of primary emotions of the living because the mechanists of depth psychology have managed to spread the opinion that all emotions spring from drives and are "therefore" irrational. This mistaken belief, so catastrophic for the well-being of life, has its irrational function and its origin in a character structure whose rationality will have to be closely examined. The primary (rational) and the secondary (irrational) emotions were found mixed together and people did not have the courage or insight to separate them. This confusion was responsible for much of the tragedy of the human animal. In order to have a complete biological understanding of this tragedy, we will have to learn far more about the function and expression of life in its natural state.

The emotions are specific functions of the living protoplasm. Living nature, in contrast to the nonliving, responds to stimuli with "movement," or "motion" = "emotion." It necessarily follows, from the functional identity of emotion and plasmatic movement, that even the most primitive flakes of protoplasm have sensations. The sensations can be understood directly from the responses to stimuli. These responses of plasmatic flakes do not differ in any way from those of highly developed organisms. There are no lines to be drawn here.

If our "impressions" of the movements of life correctly reflect their "expression"; if the basic functions of life are identical in all living matter; if sensations arise from emotions; and if emotions spring from real plasmatic movements, then our impressions must be objectively correct, provided, of course, that our sensory apparatus is neither fragmented nor armored nor otherwise disturbed.

Nonliving matter does not feel because it is without pulsatory movement. Be it a rock or a cadaver, it conveys the immediate impression of *immobility,* and with it a lack of sensation. This lack of response to stimuli is in complete accord with our impressions of inanimate matter. *Nonliving matter has no emotions,* i.e., no spontaneous movements. We will later have to go into the question of why so many human organisms "animate" the inanimate, thereby attributing sensations to it. I anticipate the principal conclusion:

Just as all emotions and reactions in life spring from and correspond to organ sensations and expressive movements; just as the living organism forms ideas of its surrounding world from impressions it derives from the expressions of the world around it; so all emotions, reactions and ideas of the armored organism are conditioned by its own state of motility and expression.

The "objective-critical" standpoint—which asserts that all perceptions of the surrounding world are "subjective," both in the unarmored and in the armored organism, and therefore "unobjective"—can be refuted by the image of an object in two different mirrors. One mirror is clear, the other has a scratched surface. The first mirror reflects the objects differently from the second. In both cases the reflection, i.e., the "sensation," is "subjective" or "arbitrary." The images of the objects in both mirrors are unreal. Yet no one can doubt that the smooth mirror will reflect the objects as they are, while the marred mirror will distort them.

I have chosen this example to show that my opponents—both the exponents of "absolute objective" science and the "critical" subjectivists—are completely correct in asserting that we have merely sensations and perceptions of the reality around us; that sensation is the only access by which the living organism is connected to the surrounding world; that we do not perceive the object itself but only its image. This is all well and good but becomes obsessional brooding unless we think further. From the standpoint of orgone biophysics, we even welcome the fact that our "objectivists" and "subjectivists" emphasize all emotional activity as being dependent on the structure of the life apparatus. It will be shown that our objectivists, the "objective natural scientists," are subjectivists, and that the subjectivists are objective observers, without knowing or even guessing it. Both claim they experience merely sensations when they describe the world. But neither asks about the nature of the sensations or, rather, about the structure of the perceiving life apparatus. Orgone biophysics has given a clear answer to this:

The unarmored being perceives the self and the surrounding world in an essentially different way than does the armored organism. Since self-awareness actually colors all other sensations and since sensation is the filter through which the world becomes manifest to us, the kind of sensations determines the kind of perception and judgments. This conclusion is indispensable and irrefutable. It applies to the unarmored being as well as to the armored—to myself as well as to my opponents, the objectivist and the subjectivist. It applies rigorously, and I am willing to abide by these terms of the debate, because the standpoint of my opponents, if fully worked through, leads to the safeguarding of functionalism, and not of mechanistics or mysticism. My position has always been that everyone is right in some way, without knowing in what way he is right.

Thus, for future reflections, we affirm the following:

The living organism perceives its environment and itself only through its sensations. On the kind of sensations depends the kind of judgments developed, the reactions based on these judgments, and the over-all picture commonly known as "world image." I did not, and do not, intend to fabricate world images. But my work and I were so often endangered by them that a closer study of their functions and foundations is in order.

The educator who thinks functionally regards the child as a living organism and shapes the environment of the child according to its vital needs.

The educator whose thinking is mechanistic and mystical regards the child as a mechanistic-chemical machine, as a subject of the state or as an adherent of this or that religion.

He presses the child into an alien world and calls this "adaptation," if he is a liberal, or "discipline," if he is an authoritarian.

That which is alive in the child obeys cosmic laws and therefore has not changed in thousands of years. It develops new meanings and contents of life from its own resources. But the adaptation to changeable life forms of mechanistic-religious civilization creates the chaos of contradictions in which the human animal finds itself trapped.

If Columbus had been a mechanistic compulsive character, he would have prepared his global circumnavigation by counting all the nails in his ship and entering them in neat columns. If, in spite of this, he had somehow reached America, he would have started the colonization by measuring the coastal stretch where he landed, by counting and measuring all trees, branches, twigs and leaves, by classifying all brooks and rivers and hills. Lost in minor details, he would have perished long before he could return to Europe to reveal his discovery of a new continent.

Our mechanistic-mystical civilization is doomed because it has filmed and classified millions of minute statistical data about the movements of a newborn infant but it still pays no heed to the biosexuality of the living organism we call child, nor to the bitter hatred that educators feel about this general fact, which looms above all others. Thus the educational question is enmeshed in details, without perspective, hopelessly snarled. The recognition of the child as a living being, instead of a future citizen, would solve all complications with one stroke because institutions would be concerned with the vital needs of the child.

Therefore, the only way out of this chaos is to shape the life forms according to the laws of the living organism. This task requires clarification about the two basically different attitudes toward life—that of the unarmored and that of the armored organism. From now on we shall operate with these two essentially different forms of life. *One is the living organism that operates, undisturbed, on the basis of natural processes. The other is the living organism whose plasmatic functions are impeded by chronic and autonomous armoring.* We expect, for good reasons, that the perceptions of the two forms are clearly distinguishable.

The unarmored organism has basically different sensory perceptions from the armored one. Since the body plasma is the receiver and transmitter of all impressions, a free-flowing plasmatic system must receive impressions that differ from the unfree, or armored one. These are not philosophical speculations about "sensation and cosmos" but hard facts, collected with great effort during many hours of daily work with the human organism; collected, organized, reexamined and finally judged in the course of more than a quarter of a century. The facts spring from concrete observations of human behavior.

The armored organism does not feel any plasmatic current, in sharp contrast to the unarmored organism. To the same degree to which the armoring is loosened, sensations of current, which the armored organism first experiences as anxiety, appear. Once the armoring is completely dissolved, orgonotic currents are experienced as pleasure. Thereby all reactions are so basically changed that we may speak of two entirely alien and essentially disparate biological conditions. Of course, this change does not occur in every instance, but where it succeeds, it parallels fundamental changes of organ sensations; and with the organ sensations, the entire view of life changes rapidly and radically.

The orgone therapist is not concerned with philosophical reflections about the world and life. Neither the physician nor the teacher, neither the patient nor the pupil broods about the "individual's attitude toward society and cosmos." I deliberately stress the clinical character of these experiences. For there are character structures that will kill off anything and everything with such phrases as: "Well, these are merely philosophical questions. The one may be just as accurate or inaccurate as the other. There are several truths about one and the same fact."

This position is untenable. *There can be only* ONE *explanation about one and the same fact that is objectively accurate; there cannot be ten different correct explanations.* The question about the energy source of biopathies admits only one answer, and not ten: *the energy of biopathic reactions stems from dammed-up biological sexual energy.* There may be several strata or phases in the development of a biopathy, different functions and aspects. There may be various ways that lead to this *one* answer. However, there is also a common feature: the basic function of energy stasis. The details may vary, depending on special social situations or childhood experiences. But what consolidates all these details, no matter how much they may differ, and is expressed basically as the biopathic deviation of the life function are invariably the stasis of biological (biosexual) energy and the armoring. In the long run, no medical science, all efforts to the contrary, will be able to evade this inescapable conclusion.

In my book *Character Analysis,* I have distinguished between two basic types, the "genital" and the "neurotic" character, with respect to biophysical "health." The fundamental difference between them is the absence or presence of a chronic sexual stasis and the autonomous armoring. The importance of this clinical distinction far transcends the individual. It deeply influences all forms of life attitudes and "world images."

The reader will no doubt have realized that what is alive in the genital character functions according to its own natural laws, while in the neurotic character it functions according to laws that correspond to the armored, and not to the living, process. A healthy snake moves and acts according to the laws of cosmic orgone energy. A snake whose body is roped up acts, perceives and responds on the basis of movements hindered by the rope. This example can be generalized:

The unarmored living organism perceives and understands the motile expressions of other unarmored organisms clearly and plainly by means of its own spontaneous movements and organ sensations. The armored living organism, however, cannot feel any organ sensations, or perceives them only in distorted form and therefore loses contact with the living process and the understanding of its functions.

For instance, an armored person may perceive his rigid, protruding chest as an expression of "toughness." Within his own life experience, this perception is correct. The "military" bearing of his chest does serve to keep his equilibrium, giving him strength in the daily struggle for existence. However, he has no idea how his natural life forces are weakened by the armoring of the chest. Furthermore, he does not understand that one can react freely and strongly with a flexible rib cage. He regards a flexible rib cage as a sign of effeminacy and weakness. He fears he cannot exist unless he "holds on to himself," instead of "yielding." He is unaware of the natural strength flowing from the free motility of the living organism. In turn, the living organism which is unarmored cannot understand how any strength can derive from a rigid rib cage. If it tries for a while to imitate the "controlled strength of character" by stopping its breath and thrusting out the rib cage, it feels only a giant effort that it cannot sustain for long. The unarmored organism cannot comprehend how the effort of the armoring can be tolerated for years on end.

What is alive in the healthy orgone therapist understands or experiences the expression of the armored patient as an effect of contrast, as it were, something alien and disturbing. He perceives the stiff rib cage or the rigid, masklike grin as embarrassing and disturbing. The armored person is different. To him, the rigid rib cage and the frozen smile have become second nature. He perceives them neither as a disturbance nor as the strain that they actually are. He does not even know that he is constantly "on guard" with his rib cage, and that his chronic grin represses tears or rage. In contradistinction, the unarmored organism perceives the fixed friendly grin for what it is, namely, an embarrassing burden to be shed as soon as possible.

Thus, the unarmored organism experiences the armoring of another organism as disturbing. By the same token, the armored organism feels the free motility of another being as alien and disturbing. In many cases, it causes conscious anxiety, and in all cases of armoring we find a deep-seated dread of free motility. Once it is loosened sufficiently, every case of biopathy shows a mortal fear of the free-moving organism. Anyone who applies the undiluted technique of orgone therapy, i.e., with the orgasm theory at its core, can convince himself that this claim is correct.

Our contention that the attitude toward life, the "world image," depends on the functioning of body plasma is unequivocally proved once we have enabled a large number of armored organisms to experience the flow of their orgonotic current. We are today in the fortunate position to give an accounting of this tool of natural research.

Basically, nature inside and outside of us is accessible to our intellect only through our sense impressions. The sense impressions are essentially organ sensations, or, to put it differently, *we grope for the world around us by organ movements* (= *plasmatic movements*). Our emotions are the answer to the impression of the world around us. Both in awareness and self-awareness, sensory impression and emotion merge to form a functional unity.

Hence, organ sensation is the most important tool of natural scientific research.

This specific claim is further affirmed by the investigations of classical natural philosophy. It asserts, with particular clarity in Kant's *Critique of Pure Reason,* that the true nature of things is not accessible to us. Observations and judgments depend on our physical organization. And, until recently, this organization was the least accessible area of nature. Therefore, the inability to recognize *"das Ding an sich"* is based, in principle, on the inability to recognize the nature of the sensory apparatus. For orgone biophysics, an extremely important conclusion follows from this accurate presupposition of the critique of pure reason:

If we succeeded in comprehending the function of perception and of sensation per se in energetic terms (orgonotically), i.e., by studying its true nature, we would create an access to "das Ding an sich." In Freud's research, it was the "unconscious" that played the role of *"das Ding an sich"* in the psychic organization and thus became the instrument of natural research. The discovery of orgone energy was made through consistent, thorough study of energy functions, first in the realm of the psyche, and later in the realm of biological functioning. The material of this research was organ sensation. It is an integral part of our ego perception as well as a part of objectively functioning nature. In orgonotic functioning, we perceive ourselves. Self-perception is objectively and experimentally controllable: if the subject of the experiment has keen organ sensations, he can, by their intensity, describe what goes on in the next room in terms of the quantitative potential fluctuations on the oscillograph. Likewise, the observer at the oscillograph can tell, by the quantity of energy variations, how intensely the subject of the experiment feels his organ sensations.[1] This removed the barrier that for thousands of years had hindered the in-

---

[1] *Cf.* Reich, *Experimentelle Ergebnisse über die electrische Funktion von Sexualität und Angst,* Oslo, 1937.

vestigation of organ sensation. X-ray photography showing the energy field of the hands was the result of the conceptual unification of sensory perception and objective excitation.[2] The energy field shows no shadows if the sensation of attraction is absent.

By closely studying the organ sensations, we become familiar with the tool we use in every kind of natural research, consciously or unconsciously, correctly or inefficiently. In this way the organization and functioning of the living organism become the most important and indispensable preconditions of scientific knowledge in general. This harmonizes with Kant's critique of reason, which places the biological organization at the root of all knowledge.

Lange drew the right conclusion early. "Perhaps," he wrote in his *History of Materialism*, "the basis for the causality concept will someday be found in the mechanism of the reflex movement and in sympathetic excitation. Then we would have translated Kant's pure reason into physiology and hence made it more graphic" (Lange, *Materialism*, Vol. II, p. 44, 1902).

Lange's grand prediction is now fulfilled. *Orgone biophysics operates with organ sensations as a first sense of a strictly physiological nature.*

In order to investigate nature, we must literally *love* the object of our investigation. In the language of orgone biophysics, we must have direct and undisturbed *orgonotic contact* with the object of our investigation. The dynamics of orgonotic body functions is revealed in orgone therapy.

In this manner we learn about the mechanisms at the root of the various life attitudes ("Weltanschauungen"). I could write a ponderous tome on the numerous armoring mechanisms to explain the mystic, the politician, the criminal, the tactician, etc. This is not the objective of this book. We must

---

[2] *Cf. Orgone Energy Bulletin,* Vol. I, No. 2, 1949.

limit ourselves to the basic mechanisms that distinguish the truly alive organism from its distorted form of expression in the biopathy. Furthermore, we have to confront these two alien worlds and penetrate to the social tragedies that have afflicted the human animal for thousands of years, ever since his organism became armored. In its own realm, the unarmored organism develops an infinite variety of life forms. The same is true for the armored organism, which develops an equally infinite variety of biopathic reactions. We are interested in the type as a whole, in the contradictory life sensations from which all other contradictions derive.

The armored organism is essentially different from the unarmored one in that a rigid wall is erected between its biological core, from which all natural impulses stem, and the world in which it lives and works. As a result every natural impulse, particularly the natural function of and capacity for love, is impeded. The living core of the armored organism has retained its impulses, but they can no longer find free expression. *In the desperate attempt "to express itself," every natural impulse is forced to penetrate or break through the wall of the armoring. The impulse must use force to reach the surface and the goal.* While the impulse is trying to overcome the armoring by force, it is transformed into a destructive rage, regardless of its original nature. It does not matter what happens to this secondary rage reaction later, after passing through the armor; whether it spends itself or is inhibited, whether it turns into morbid self-pity or reaches its goal as undisguised sadism: *the core of the process is the transformation of all love impulses into destructiveness while passing through the armor.* To repeat: it is the effort to express itself naturally and reach its goal that converts every basic biological impulse into destructiveness. In the process, the total being of the armored person acquires a characteristic that can only be described as hardness or disharmony.

The diagram of this process is as follows:

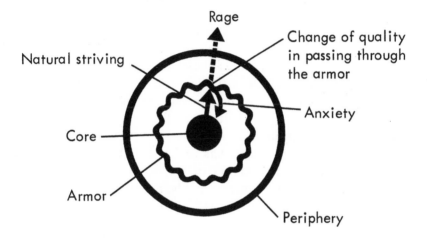

The existence of the armor does not prevent the stricken organism from loving or being afraid. Its life expressions spring from all depth strata of the organism. It communicates with the world as if through gaps or holes in the armor. But since it cannot fully let itself go, its love is small, carefully measured and allotted; its concern for the child is "controlled," "taking all circumstances into account"; its accomplishment is "well-measured" and "reasoned," pointing toward meaningful and "purposeful" work; its hatred is "goal-oriented" and "circumscribed." Briefly, it never loses its head, is always "reasonable" and "composed," the way a "realistic politician" has to be. Such an organism hates the well-ordered but infinitely variable freedom of natural processes, or it fears them.

Its destructive hatred is directed mainly—we would not exaggerate if we said, exclusively—against all genuine and unrestrained attitudes of the unarmored organism, against the spontaneous, pleasurable, enthusiastic, vibrating, wild and foolish things in life. Above all, it opposes what is involuntary and free in the body. In its destructive attitude toward the alive or unarmored organism, the armored organism knows no mercy. Here it loses the qualities it has otherwise raised to the level of ideal human behavior. In the guise of idealistic or hygienic attitudes, the armored organism knows how to kill every spontaneous impulse in itself and other organisms. One should carefully read the following piece of advice, printed in a widely read New York newspaper in 1945. The author has realized the importance of proper breathing. She knows of the harm done by poor breathing. She wants to correct and help, but what a regimen of "you may" and "you must not" is unfolded!

DEEP BREATHING EXERCISES

The effects of deep breathing are as varied as they are beneficial. The blood is purified, the complexion improved, and adequate assimilation of food is aided when we breathe correctly.

There are some valuable exercises which improve posture, promote the health of internal organs, and encourage correct breathing habits. You may want to add these to your daily routine.

No. 1: Lie on the floor, on your back. Place your right hand on your diaphragm. Inhale slowly and deeply. As you inhale, your diaphragm and abdominal wall should move upward toward the ceiling. As you exhale, the area under your hand should return to normal position. Then contract abdominal muscles and pull in, and at the same time, flatten your back against the floor, touching as much of it as possible to the floor.

No. 2: Remain in the same position. Inhale deeply and bend your knees, placing your feet flat on the floor. Begin inhalation at the diaphragm so you feel the ribs spread sideways. Do not lift your upper chest, however. Exhale by drawing your abdominal muscles in and up, but do not allow ribs to collapse. Take another breath and exhale. Now take a series of little breaths, one right after the other. Exhale in a series of small exhalations by contracting the abdominal muscles in quick exhalations. Complete by entirely exhaling and collapsing the rib cage.

This person recommends precisely what our armored patient tries to do in a biopathic manner when we ask him not to hold his breath. He "does," "exercises," "demonstrates," "acts," and "performs breathing." But he does not breathe according to natural innervation. We can safely apply this example to all life expressions of the armored organism. Its nature invariably consists in evading and impeding the spontaneous, direct expression. In all other life situations the armored person may be tolerant, even charming, friendly and helpful. He only becomes stubbornly angry when he sees the living organism functioning without armor.

The destructiveness of armored life against unarmored life may be observed in the relations of most educators toward newborn children. The newborn infant comes into the world without armor. Life functions in him without regard for the "demands of culture." His first utterance activates his orgonotically highly charged mouth. In our highly reputable obstetrical hospitals, infants, during their first twenty-four to forty-eight hours, are not breast-fed by their mothers, according to some iron law. It takes a threat to move a nurse or a physician to break this rule. The infants suffer and whimper. "Culture" has no ear for that. One might ask about the reason for this procedure. There is no reasonable answer, or at best one of those stereotyped replies that might come from a masked mouth. Newborn infants feel the touch of their mothers for only a few minutes during the day. Think of it: what an offense against the rules of "hygiene"! The infant, barely torn from the orgonotic contact with the warm uterus, which lasted for an

uninterrupted nine months; this infant, suddenly transferred from an environment of 37° C. into one of 18 or 20°
C., may not feel the mother's body. This violates the rules
of hospital administration, culture and morality; it provokes
the Oedipus complex, offends customs and mores, opposes
the life style of the highest of all branches of medicine, represented by the academies of science, honorary doctors and
honorary presidents of all universities of this universe, in
which electrons and protons, side by side with neutrons and
positrons, dance the St. Vitus dance of atomic explosion. It
is here, precisely here and not at diplomatic conferences or
anywhere else, that infants acquire their future readiness to
make war. The newborn child reacts to the cold first with
anxiety, then with screams, and finally with a contraction of
his autonomic system, the first contraction of his life, unless
a lifeless uterus has damaged his organism beforehand.

This massacre of the newborn, plainly audible in the ear-
splitting and heart-breaking screams in all the infant rooms
of all obstetrical clinics the world over—this massacre, as I
said, has nothing to do with hygienic considerations. It is the
first unconscious but drastic measure by armored organisms
in the guise of physicians, administrators and parents
against the living organism that confronts them, unspoiled
and undistorted. Let us ponder this fact: thousands of doctors and nurses hear the screaming infants and understand
nothing. They remain deaf and dumb. Add to this the
ritual of covered mouths and rubber-clad hands to understand the shifting of all emphasis, the evasion of the essential, and the stress on the unessential, which are used in this
battle against the newborn child.

I insist that prevailing medicine and pedagogy, as they
are officially taught and practiced, do not understand the living organism and do not have the vaguest idea about the most
primitive life processes. I speak of life processes, and not of
the number of red blood corpuscles. This confirms my rigor-

ous dictum that *the organism can perceive only what it itself expresses.* The armored physician cannot hear the screams of the infants, or else he takes them for granted because he has stifled the screams inside himself; and because his own organism can no longer perceive what another organism reveals to him. There are only a few, courageous islands of inchoate understanding.

I do not intend to write a compendium of the massacre committed in the upbringing of the child. This must be left to others who are more competent than I am. First of all, the common denominator of its many variations must be brought to light. I do not doubt the good intentions of doctors, nurses, educators, and parents. I do assert, however, that their deep-seated love for the newborn cannot be translated into practical action. They do not understand the living function in the newborn; more than that, they fear it as dangerous and alien. Psychoanalytic pedagogy has openly proclaimed the "killing of the animal" in the child. This attitude deserves sharpest condemnation and the revelation of the motives behind it.

Loss of contact with the maternal organism is the first step in the following sequence: impeding of any kind of motor activity (a mere few years ago in the form of actually binding limbs and torso by "swaddling"); forcing the infant to breathe through the nose instead of breathing naturally through the slightly opened mouth; rigid upbringing to suppress crying and screaming; regulation of the bowel movements and forced toilet training with suppositories and enemas by compulsive mothers; prevention of all spontaneous sexual expressions, especially masturbation.

Armored life encounters unarmored life with anxiety and hatred. In many cases, the anxiety turns into shock, and the hatred into bloodthirsty, murderous and blindly raging terror. The armored organism cannot bear the soft, yielding movements of a newborn child. Anyone who has ever held a bird in his hand knows the feeling I mean. It is not easy to understand why the language of our educators is so sharp, their demands so stern, their tone so harsh, their punishments so calculated to humiliate. In the conceptual realm of present-day culture we find no satisfactory explanation that sheds any light on this rough treatment of children. There are no plausible reasons that explain why the ideal teacher is still the elderly, gaunt, ugly, virginal spinster. Celibacy for teachers is still practiced in many countries, and where it is not mandatory it is tacitly expected.

It has long been known that the sexually unsatisfied organism experiences natural life expressions as a provocation. It is known that such a person does not like to be reminded of his unhappiness, of his frustration. But it is not that simple. The armored organism does not initially hate the truly alive, unarmored organism. On the contrary, he tries to establish contact and reacts first in a rational, loving manner. But the course of the love relationship invariably and compulsively turns to furious hatred. Close examination of the process shows that the armored organism either cannot fully establish contact or cannot maintain it. Sooner or later the warm, loving impulse is blocked by the armor. This is followed by the tormenting feeling of frustration. *The armored organism does not know that its love impulses have been blocked, and behaves as if the unarmored organism had denied it love.* In its desperate attempt to break through and express its love, the love impulse turns into hatred and destructiveness. This hatred is not willed, not consciously reasoned. The reason for the hatred is always dragged in as a by-product, thus becoming a rationalization.

The mechanism just described has general importance. Without recognizing it, so much of human interraction, which otherwise would be incomprehensible, can be understood. In another context, I will discuss the lawfulness of this sudden turn from rapture and love to bitter hatred.

It is always the inability of the armored organism to develop its love impulses, to generate enthusiasm, to commit itself without reservation to a cause that makes it destructive.

In addition, there is the attitude of the unarmored child toward the armored educator. Healthy children are extremely sensitive to simulated forms of behavior. They turn away from armored people and reach out to those who are unarmored. In the presence of an armored organism, the unarmored person very soon feels "hemmed in," "without contact," "rejected."

We are not yet able to read reactions accurately with the use of instruments, but there can be no doubt about the biophysical nature of these reactions. We are dealing with contact phenomena, with excitations of the orgone field.

The contact between two unarmored organisms—say, between a healthy mother and a healthy child—is completely different. There is nothing forced about body expressions. Love attitudes are just as rational in their foundation and expression as are defensive mechanisms or hatred. Contrary to the complicated and tangled conflicts in the relationship between two armored organisms, healthy human intercourse is simple. It is not without friction, but there are no biopathic reactions of neurotic stickiness, petty vindictiveness or insoluble complexities.

It is precisely this "simplicity" in human intercourse that is so incomprehensible to the armored organism. *Everything natural and profound is simple.* The simple, grand lines of emotional expression are known to characterize the great painter, musician, poet, novelist and scientist. But the simple is alien to the armored organism. Its impulses are so complicated in their form of expression, the manner of their utterance is so muddled and contradictory that it has no organ for the simple and unequivocal emotional expression. It even lacks a sense of simplicity. Its love is mixed with hatred and anxiety. The unarmored organism loves unequivocally in love situations, hates unequivocally where hatred is legitimate, and fears unequivocally where fear is rational. The armored organism hates where it should love, loves where it should hate, and is frightened where it should love or hate. *Complexity is the specific life expression of the armored person.* He is trapped, as it were, in the multiple contradictions of his existence. Since he approaches all experiences with his complex character structure, his experiences become equally complicated. He is amazed at the accomplishments of the healthy organism, or he places these accomplishments in the area of special talent barred to him. "Genius" becomes a kind of abnormal monster, because he cannot understand the great simplicity in the life expression of "genius." In the consistent stripping away of the layers of character, one discovers that complexity epitomizes the defensive mechanism in its purest form. The armored person is complicated because he has a mortal terror of everything simple, straightforward and direct. I say: *mortal terror.* This is no literary exaggeration. The word accurately describes the process: *the simple, straightforward, direct expression inescapably leads periodically to orgastic plasma convulsions.*

The armored person cannot express himself with immediacy because his natural impulses are distorted, fragmented, inhibited, and transformed in the tangled net of his character structure. The armored person perceives himself and the world as complicated because he has no immediate contact, no straightforward relationship to the world around him. The secondary result is that over the years this world becomes actually complicated. Since these complications make an ordered existence impossible, artificial rules in "human intercourse" emerge: rigid mores, customs, rules of etiquette, diplomatic maneuvers.

No unarmored person would have the impulse to vomit or pass wind in the presence of others. Hence it would never occur to him to advise or demand obedience to the rules of society that one must not vomit in the presence of others or yield to flatulence. The armored person, however, is full of such impulses. It is from his own source of secondary drives that he derives the rules and ethical customs against them. The nature and intensity of asocial secondary impulses in man can be directly established by the number and strength of societal mores and rules of conduct.

The unarmored organism does not know the impulse to rape or murder young girls, to experience pleasure by violence, etc. It is therefore indifferent toward moral rules designed to hold these impulses in check. It finds it inconceivable to have sexual intercourse solely because there is an opportunity, such as being alone in a room with a person of the opposite sex. The armored person, on the other hand, cannot imagine an orderly life without rigid compulsory laws against rape and sex murder. He is incapable of understanding that a man and woman can be alone without sexual intimacy. Briefly, he is full of perversions, and therefore the world appears to him as one vast perverse temptation. For a long time, psychoanalysis had to suffer from this perversion of the perverse as it raged against the physician and the female patient being alone in a room. Even today, sex economy is regarded as a doctrine of sexual orgies among perverse and armored persons. A heavily armored sadist who came to me for help was enthusiastic about my writings because I had, so he said, "completely set free the indiscriminate fucking of everybody with everybody else." After several weeks of hard work, he knew better. I have no control over the many perverts and sadists who read my books and pass off their dirty fantasies as my doctrine. He who fights the plague must reckon with the possibility of becoming infected himself. The pornographic interpretation of sex economy is the infectious disease of our profession.

In time, we will convince ourselves that this pornographic

infection is not limited to the circle of pornographers and perverts. Extremely "honorable" groups contain equally dangerous carriers of the emotional plague. One of the objectives of this book is to distinguish sharply the healthy from the sick life, to implant their contradictions into human consciousness and thus protect the physician who fights the emotional plague from such dangerous "contamination."

Sadistic brutality, pornographic lewdness and rejection filled with mortal terror are thus the external manifestations of the impulses after they have passed through the character armor. Since the majority of human animals are armored, it is not surprising that sex economy and, later, orgone biophysics ran into a roadblock of brutality, pornography and mortal terror after they had made some simple, directly understandable disclosures concerning the core of the life process. If the responsibility for separating natural life impulses from the sadistic pornography of the general public were not so enormous, I would have declined any controversy with those on a giant garbage heap of emotional filth. As it is, I could not withdraw into the comfortable academic viewpoint of splendid isolation. There is too much at stake. And while not motivated by vindictiveness, I cannot deny myself the triumph of having escaped unbroken and untainted, from this truly Dante-esque inferno of human existence, even though badly scarred by painful and dangerous insults to my sense of dignity and awareness of my scientific accomplishment by the outrageous slander of biopathic individuals.

From now on we shall trace the unarmored living organism from its origins in the cosmic orgone ocean. And we shall follow the armored organism on its grotesque and perilous road. We will separate these two basic forms of life and see them collide only at specific places. This separation is artificial, and like everything artificial it is only a dead cross-section through living functions. We will leave important questions unanswered, and many doubts will remain. But, at the end, a law will be revealed that brings all orgonotic functions in their infinite shapes and varieties into a unified, grand harmony of nature. Man's deviation from this law will be all the more painfully experienced. What I have grasped and described in medical language as the EMOTIONAL PLAGUE OF THE HUMAN ANIMAL will strike us as even more terrible. It constitutes the vast disharmony in human nature, alien to the animal and the plant, the human infant, and the few human beings fortunate or strong enough to escape it.

In describing the deadly struggle between unarmored and armored life, I can report only on those cases I experienced personally or as a physician with my patients. In doing so, the limits between orthodox biology and sociology tend to blur. Man as a social being is only a specific variant of his natural existence. In the light of the harmonious process of life, the raging of the emotional plague in social life will reveal itself glaringly. We will have to transcend the realm of our social thinking and look at it from the standpoint of the living process in man in order to gain distance and perspective. We will have to assess the great tragedy that befell the human animal when it viewed what is alive in itself: its worst enemy. We will experience its hatred of the living. But only if we look upon the social realm of functioning from the standpoint of the living process will it be possible to condemn not plague-stricken man but the circumstances that degraded him into the most malevolent of all animals.

We are setting out on a long and arduous journey. Those who do not want to come along or who expect the usual platitudes about biology and sociology should stay home. But those who have the courage to face some hard truths are welcome to join us.

CHAPTER IV

ANIMISM, MYSTICISM, AND MECHANISTICS

We must ask: was ignorance about the living process merely the result of a faulty conceptual technique and insufficient research? Or was it the result of a characterological inhibition, an unconscious intention, as it were? The history of science leaves no doubt that the living process was not allowed to be studied; that through thousands of years it was the mechanistic-mystical structure of the human animal that excluded from all research, by absolutely all conceivable means, the cosmic foundations of the living process. And there was method to this structural intention: first, in the religious conceptual prohibition that presented God and life a priori as being unknowable; then, even in the punishment by death for recognizing the life process, as contained in several religious taboos. The myth of Adam and Eve has a deep rational meaning. To eat from the tree of knowledge meant to be expelled from paradise by fire and sword. It is a snake, a symbol of the phallus as well as of primal biological motion, that persuades Eve to seduce Adam. They pluck the forbidden apple and eat it. They are overcome with shame. The sexual symbolism is self-evident: "whoever eats from the tree of knowledge perceives God and life, and this is punishable"—thus goes the legend. *Awareness of the law of love leads to awareness of the law of life, which in turn leads to awareness of God.* This sequence is true throughout every phase of the history of science and has been confirmed by the discovery of cosmic orgone energy in the twentieth century. The punishments that followed this discovery were entirely in keeping with the old Biblical legend.

It is not true that I was the first to observe orgone energy and, with it, the functional law that merges organic and inorganic nature into one. In the course of two millennia of human history, people time and again encountered phenomena of orgone energy, or they developed thought systems that approached the reality of cosmic orgone energy. That these insights could not break through should be blamed on the same human character traits that created religious prohibitions and destroyed any progress in the right direction. Basically, the weapons of destruction were invariably either mechanistic, pseudo-scientific counter-arguments or mystical obscurantism, except in cases of physical annihilation.

I can only cite scattered examples from various epochs. A presentation of the systematic effort on the part of the emotional plague of mankind to destroy the functional equation of God = life = cosmic orgone energy = orgonomic functional law of nature = law of gravitation must be left to historians.

Even among the Greeks there existed a rigid and fanatical orthodoxy that rested just as much on the interests of an arrogant priesthood as it did on the faith of a multitude in need of redemption. This might have been forgotten altogether if Socrates had not been forced to drink the cup of poison; but Aristotle, too, escaped from Athens so that the city did not sin against philosophy a second time. ["Philosophy" in antiquity played the role of today's natural science—W.R.] Protagoras had to flee and his writings about the gods were burned by order of the state. Anaxagoras was imprisoned and had to flee. Theodorus the atheist and probably Diogenes of Apollonia were persecuted as atheists. And all this happened in humane Athens. From the standpoint of the multitude, everyone, even the most idealistic philosopher, could be persecuted as an atheist, for no one imagined the gods to be as priestly tradition dictated it.

This was written by Friedrich Albert Lange in his *History of Materialism*. What was the basis for the ancient denial of the gods? What was there about the development of scientific materialism that enabled Greek philosophy to oppose superstition? It was the energy hypothesis of the "soul atoms" of Democritus, i.e., a scientific glimpse of the existence of a special energy, the orgone, underlying the psychic functions.

Materialistic philosophy did not start from mechanistic questions but, oddly enough, from basic psychological questions, just as orgone biophysics sprang from the psychiatric problems of the biological drive dynamics: WHAT IS SENSATION? HOW CAN MATTER PERCEIVE ITSELF? WHAT IS SENSATION TIED TO? UNDER WHAT CONDITIONS DOES SENSATION EXIST, AND UNDER WHAT CONDITIONS DOES IT NOT?

Thus, natural science of antiquity, brilliant and in its assessments accurately oriented to this day, did not start from material but from functional problems that did not exclude sensation. It was over these functional processes, and not over materialistic questions, that the scientific minds separated from the metaphysicians and the mystics. These were the questions that kindled the battle flames of the emotional plague against the knowledge of nature and the equation of God = natural law. These questions—and not, originally, the mechanistic laws governing the velocity of falling bodies —turned the accurate world image and its processes into the burning issue they became. For every scientific mind realized that it was only our own sensation of the processes in us and outside us that contained the key to the deeper secrets of nature. The sensation of living protoplasm is a singular phenomenon of nature, within and not beyond human life.

Sensation is the sieve through which the inner and outer stimuli are perceived; it forms the bridge between ego and outer world. This is an established fact, both among natural philosophers and natural scientists who are aware of their investigative methods. It is all the more strange that, until recently, scientific research was unable to say anything about this central part of its own nature and that mysticism could usurp the realm of life sensations with such completely disastrous effect.

Such grotesque facts always have a certain function and a secret intention. It would of course be wrong to assume that sometime, somewhere, there was a secret council of armored human animals who determined how the knowledge of sensation itself—the link between ego and nature—could be prevented and why those who discovered this secret of nature should be so harshly punished, persecuted, burned and tortured. There were no secret deliberations and there were no decrees. The deadly battle that the emotional plague waged against the recognition of the basis of sensation was dictated, guided, and carried out by the structural laws of armored human beings.

It was the character-analytic theory of structure that first broke the spell and opened the gate to insight into the nature of sensation itself. The ensuing discovery of biological energy phenomena within the perceiving organism and the further discovery of atmospheric orgone energy in the purely physical sense were merely the logical consequences of the first act: *the discovery that sensation is a function of excitation; that, in other words, there is a functional identity between the quantity of excitation and the intensity of sensation.* With this, sensation itself had become the objective of scientific research. The further consequences of this discovery speak for themselves.

Sensation is a function—the function of a limiting membrane that separates the living system from the surrounding orgone ocean. Through this membrane the orgonotic living body communicates with all other orgone systems. It is no accident that the sensory nerves develop from the ectoderm, the outer germinal layer of the gastrula.

Since the physical view of nature results from the biological constitution of the observer of nature, the world image cannot be separated from the creator of the world image. Briefly, natural research that discovered the atom bomb is confronted with the natural research that discovered cosmic orgone energy, sharply, clearly, and irreconcilably.

It must be decided whether nature is an "empty space with a few widely scattered specks" or whether it is a space full of cosmic primordial energy, a continuum that functions dynamically and obeys a generally valid law of nature.

The technician of physics, whose thinking was shaped by mechanistic philosophy, regards absolutely all physical problems as basically solved. His view of life is limited by the fact that sophisticated equipment enables airplanes to fly by remote control, so that the living pilot is eliminated. He believes that the invention of the most infamous murder weapon since time began is "the dawn of a new era of atomic energy." His universe is crumbling under his feet, but his view of life is fixed and compact, consisting primarily of an "empty space" in which there are "a few specks." We do not want to argue with him, although his opinion is an essential part in forming general public opinion. There is no room for the life process in this view. More than that, the practical effect of his view of nature is destructive: in theory, by omitting the living substance from all consideration; in practice, by social murder and war.

It is different with the founders of this dead and deadly philosophy. The founders of this empty and dead universe are intelligent, educated men. They do not believe that all problems have been solved. On the contrary, they openly say that their physical image of the world is urgently in need of correction. They find themselves in contradiction with their own theory. In their own words, they have abandoned reality to withdraw into an ivory tower of mathematical symbols. They cannot be blamed for withdrawing from the real world into a shadow world and for operating with shadowy things and abstract symbols. Everyone can do, or not do, as he pleases, provided he harms no one. But does this kind of influence really harm no one? Is not the damage proved, since this kind of physics excluded the human being, mysticized life, and, intentionally or not, invariably returned to explosive substances because of its research orientation?

I will try to describe the sensory apparatus of the mechanistic observer who has created mechanistic philosophy. How does it happen, we must ask, that mechanistic physics has been declared bankrupt by its most brilliant exponents but so far has failed to break through the iron walls of the thinking in which it is trapped? If we are consistent and hold the character structure of the physicist responsible for the mechanistic view of life, we must ask these questions: What is the nature of the mechanistic character structure? Which specific qualities underlie its helplessness in observing nature? Where does this character structure come from? And, finally, in which social processes did it originate?

I am not obliged to present the history of mechanistic natural science. Suffice it for me to speak from experience and to describe the typical mechanistic physicist as he is revealed by psychiatric investigation.

The typical mechanistic physicist thinks according to the principles of machine construction, which he primarily serves. A machine must be perfect. Hence the thought and action of the physicist must be "perfect." *Perfectionism is an essential characteristic of mechanistic thinking.* It permits no errors. Uncertainties, situations in flux, are undesirable. The mechanist works with artificial models of nature when he experiments. The mechanistic experiment of the twentieth

century has lost the essential features of genuine investigation
—the control and imitation of natural processes, which have
characterized the work of all pioneers in the natural sciences. All machines of the same type are alike down to the
most minute detail. Deviations are regarded as inaccuracies.
In the realm of machine construction this is quite correct. But
this principle will lead to error if applied to processes of nature. Nature is imprecise. Nature does not operate mechanically but functionally. Therefore, the mechanist always goes
against nature whenever he uses his mechanistic principles.
There is a lawful harmony of natural functions that permeates and governs all being. But this harmony and lawfulness
is not the mechanical straitjacket that mechanistic man
has imposed on his character and his civilization. Mechanistic civilization is a deviation from the law of nature; even
more, it is a perversion of nature, an extremely dangerous
variant, just as a rabid dog represents a morbid variant
within the species.

In spite of the lawfulness of their functions, natural processes are characterized by the absence of any kind of perfectionism. In a naturally grown forest we find a uniform
principle of growth. But there are no two trees—and no two
leaves among the hundreds of thousands of trees—that resemble one another with photographic likeness. The realm
of variations is infinitely wider than the realm of uniformity.
Although the uniform law of nature can be found and functions in every single detail, no matter how small, there is
nothing that can be reduced to perfectionism. With all their
lawfulness, natural processes are uncertain. Perfectionism
and uncertainty are mutually exclusive. One cannot object to
this fact by pointing to the certainty of the functions of our
solar system. True, for thousands of years the orbits of the
planets around the sun have not changed. But thousands,
even millions, of years play only a minor role in the processes of nature. The origin of the planetary system is just as

uncertain as its future. This is generally recognized. Thus, even the planetary system, this "perfect" mechanism of the astrophysicists, is imperfect, in the "irregular" fluctuations of thermal periods, sun spots, earthquakes, etc. Neither weather formation nor tidal flow and ebb functions according to the laws of machines. The failure of scientific mechanistics in these realms of nature is obvious, as is their dependence on the functions of a primordial cosmic energy. There is law in nature; that much is certain. But this law is not mechanistic.

Therefore, perfectionism is a compulsive accuracy of mechanistic civilization; accurate within but not outside the realm of mechanistic functions, the artificial models of nature. Just as everything within the conceptual framework of formal logic is logical, but becomes illogical outside this framework; just as everything within the framework of abstract mathematics is consistent, but outside has no frame of reference; just as all principles operating in the authoritarian educational system are logical, but outside are useless and anti-educational; so, too, is mechanistic perfectionism outside its own logical domain unscientific; and, in its pseudo-accuracy, it functions as a drag upon natural investigation. Research without errors is impossible. All natural research is, and always was, groping, "irregular," unstable, flexible, forever corrective, in flux, uncertain and insecure, and yet in contact with real processes. For these real processes, in spite of all their basic unifying laws, are variable in the highest degree, free in the sense of being irregular, unpredictable, and unrepeatable.

It is precisely this freedom found in nature that frightens our mechanists when they encounter it. The mechanist cannot tolerate uncertainty. But this freedom is neither metaphysical nor mystical but functionally lawful.

Here character analysis has opened up several crucial insights. It was important to apply psychiatric insights into human reactions, to the basically incomprehensible, hate-filled rejection of orgonomic phenomena. In my publications, I have dwelt time and again upon the astonishing fact that cosmic orgone energy was so thoroughly and so consistently overlooked by the physicists. Lasting as it did for centuries, this oversight could not be an accident. My psychiatric work fortunately enabled me to unravel part of the mystery during the character analysis of an extremely gifted but inhibited physicist of the classic mechanistic school.

We found that, rooted in certain experiences irrelevant in this context, he had developed a strong, dreamy cosmic longing and fantasizing from childhood on. It had led him to study physics. The core of this fantasy was the idea of floating all alone and lonely in the universe among the stars. A specific memory from his second year showed that this fantasy was rooted in real personal history. As a small child he had observed the stars at night through the window. He waited for their appearance with an excitement mixed with anxiety. His "flight into space" also served to remove him from very unpleasurable situations in his parental home. The strong inhibition I mentioned earlier sprang from precisely those painful experiences that led him to study the universe. But, at the same time, they remained as permanent inhibitions on his capacity for surrender and as a block to his organ sensations. As we approached the liberation of his orgastic sensations, a severe anxiety emerged, an anxiety that was at the heart of his work block. It was the same anxiety he had developed as a child due to his powerful organ sensations. In organ sensation, man experiences the orgone function of nature in his own body. Now this function was charged with anxiety and therefore inhibited. Our physicist wanted to devote himself to orgone biophysics because he was convinced of its accuracy and significance. He had seen the orgone in the metal room and described it in detail. But whenever he was supposed to do practical work, a strong inhibition set in, the same inhibition based on the fear of total surrender, of unquestioning abandon to his own body sensations. In the process of orgone therapy, the sequence of moving forward and fearfully retreating was repeated so often and so typically that there could be no doubt that *the fear of organ sensations and the fear of scientific orgone research were identical.*

The reactions of hatred that came to the surface were the same that one encounters in ordinary relations with physicists and physicians regarding the orgone. Our clinical experience may be generalized: *it is the fear of autonomic organ sensations that blocks the capacity to observe orgone energy.*

Self-perception is the deepest and most difficult problem of all natural science. The understanding of sensation will also pave the way for understanding self-perception. We recognize the capacity for sensation in living organisms by their response to stimuli. This response is inseparably connected with an EMOTION; in other words, with the *motion of protoplasm.* We know that an organism has perceived the stimulus when it responds with movement. Emotional stimulus response is functionally identical with sensation, not only quantitatively but also qualitatively. Just as all stimuli affecting an organism can be reduced to two basic forms—pleasurable and unpleasurable—all sensations can be basically reduced to two fundamental emotions, *pleasure* and *unpleasure.* This fact was already known to pre-Freudian psychology; it was clarified by Freud with his libido theory. The accomplishment of orgone biophysics was that it succeeded in functionally equating *pleasure* with biological *expansion,* and *unpleasure* (or anxiety) with biological *contraction.*

Expansion and contraction are basically *physical* functions that can be found even in the inorganic realm of nature. They comprise much wider areas than emotions. It may be assumed that there is no emotion without expansion or contraction, but that expansion and contraction function without emotion, as, for instance, in the atmospheric orgone. We reach the conclusion that the emotions, at a certain point in the development of living matter, are added to orgonotic expansion and contraction when certain conditions are fulfilled. For the moment we assume that *emotion is tied to the existence and movement of protoplasmatic substance within a circumscribed system and, without this precondition, does not exist.* But in another context we will encounter the susceptibility of the purely physical orgone to stimuli when we discuss the medium of electromagnetic waves.

Many problems are still awaiting concrete answers. But regardless of the many obscurities blurring our vision, it is certain that from now on sensation and emotion are found within, and no longer outside, the physical view of nature. Mechanistic natural research must exclude sensation because it cannot grasp it. But since sensation and emotion are the direct and least doubtful experience of the living organism, they were bound first to strike the attention of the natural philosophy of antiquity and to press for an answer. In his book *Meeting of East and West,* Northrop explains the importance of direct organ sensation for the entire natural philosophy among ancient Asiatic cultures. It was not ascribed to some god. It was treated within the framework of physical functions and attributed to special, particularly smooth and exceptional, atoms. This ancient view is far superior to that of "modern" natural science and comes closer to the natural processes.

The primitive view of emotional life was not mystical, as is our view today; neither was it spiritualistic or metaphysical. It was *animistic.* Nature was regarded as "animated," but this animation was derived from man's own real sensations and experiences. The spirits had human form, the sun and the stars acted like real, living people. The souls of the dead continued to live in real animals. The primitive animistic intellect did not change the world within or without. The only thing it did in contradiction to natural scientific philosophy was to ascribe real functions to real objects where they did not belong. It placed its own reality into an alien reality, that is, it projected. The primitive intellect reasoned very close to probability when it equated the fertility of the earth with the fertility of the female body, or when it regarded the rain-bearing cloud as a being capable of perception. Primitive man animated nature according to his own sensations and functions; he *animated* them, but he did not mysticize them, as did his successor several hundred years later. *"Mysticism" here means, in the literal sense, a change of sensory impressions and organ sensations into something unreal and beyond this world.* Anthropology teaches us that the devil with tail and pitchfork, or the angel with wings, is a late product of human imagination, not patterned on reality but originating from a distorted concept of reality. Both "devil" and "angel" correspond to human structural sensations that deviate basically from those of animals or primitive men. Likewise, "hell" and "heaven," formless, blue-gray ghosts, dangerous monsters and tiny pygmies are projections of unnatural, distorted organ sensations.

The process of animating the surrounding world is the same with the animistic primitive as it is with the mystic. Both animate nature by projecting their body sensations. *The difference between animism and mysticism is that the former projects natural, undistorted organ sensations, while the latter projects unnatural, perverted ones.* In both cases we can draw conclusions about the emotional structure of the organism from mythology. But we can also discern the radical difference. It forms the transition from a biological form of existence to a basically disparate life form of the human animal.

We can still describe animism as a realistic conception of nature, even if the animating idea and the animated object do not concur in reality. For both idea and object are objective, unchanged realities. But we cannot regard mysticism as a true conception of nature because not only the outer but also the individual inner world has deviated from the law of nature; they have changed. Animism takes for granted a soul in a cloud or the sun, which is not correct, but it does not tamper with the form and function of such natural objects. A devil or an angel, on the other hand, no longer corresponds to any reality, neither in form nor in function. The only reality at the root of this mystic kind of animation is the distorted organ sensation of armored man.

This discussion is of decisive importance in clarifying several basic questions of natural research. Later we shall see that Kepler had an animistic concept of planetary functions which we should not confuse with mysticism, although he has often been accused of it. Furthermore, it is known that Galileo, who established the mechanistic functional laws, was not on good terms with Kepler. We will also find an animism in Newton that we will have to understand. It is important to dissociate ourselves from the unfounded superiority of the mechanists, who dismissed as "mysticism" the animistic efforts of a Kepler or Newton to comprehend the harmonic law of nature. We will have to demonstrate that our mechanists are far more mystical than they themselves suspect, and far more remote from nature than primitive animists. History shows that mechanistics in the natural sciences developed not as a reaction against the animism of a Democritus or a Kepler but against the rampant mysticism of the Church in the Middle Ages. The Christian church had exchanged the nature-oriented animism of prehistoric science and the vitality of its own founder for a mysticism remote from life and nature. The mystical bishop sent the animistic "witch" to be burned as a heretic. Tyl Ulenspiegel was a nature-oriented animist; Philip II of Spain was a sadistic-brutal mystic. Functional natural science must defend primitive animism against perverse mysticism and take from it all elements of experience corresponding to natural sensory perceptions.

Narrow-minded mechanists of natural science reproach functionalism for being "mystical." This blame rests on the assumption that those who try to understand mysticism are mystical. The mechanist does not understand emotional processes at all; to him, they are alien experiences, and they are equally foreign as the object of investigation. In a manual of neurology or organ pathology one will look in vain for a study of the emotions. The emotions, however, are the experiential material of mysticism. Therefore, the narrow-minded mechanist concludes that those who deal with emotions are mystics. The understanding of emotions is so remote to the mechanist's thinking that there is no room for it in his natural scientific investigation. Functionalism is simply not capable of overlooking the emotions and can include them in the realm of natural scientific research. The mechanist regards this as "mysticism" because they confuse mysticism with the study of mysticism.

For mechanistic pathology, functional illnesses are "imaginary" illnesses. When the mechanistic physician cannot ascertain any alteration in the chemical composition of the blood or in the structure of the tissues, he cannot diagnose an illness, even if the patient actually dies. The functionally oriented physician, the orgone therapist, knows about the bodily function of the emotions. He understands how and why one can "die of grief." For grief is functionally identical with the shrinking of the autonomic nervous system, a protracted shock, as it were. For him, "functional fever" is not a figment of the imagination but a real, biophysically interpretable excitation of the biosystem.

The distinction between animism and mysticism is important insofar as the orgone-physical motility of living substance can be differentiated from the animation of lifeless substance (= animism) and the grotesque distortion of organ sensations (= mysticism). For the mystic, a soul "lives" in the body. There is no connection between body and soul except for the fact that the soul influences the body, and vice versa. To the mystic (and to the mechanist, if he is aware of any emotional factors at all), body and soul are rigidly separated though interrelated realms. This is true for both psychophysical parallelism and mechanistic and psychologistic causal relationship: body → psyche, or psyche → body. Functional identity as a research principle of orgonomic functionalism is nowhere as brilliantly expressed as in the unity of psyche and soma, of emotion and excitation, of sensation and stimulus. This unity or identity as the basic principle of the concept of life excludes once and for all any transcendentalism or even autonomy of the emotions. Emotion and sensation are, and remain, bound to the orgone-physical excitation. This also excludes any mysticism. For the essence of mysticism lies in the concept of a supernatural autonomy of emotions and sensations. Hence, every concept of nature that is based on the autonomy of emotions, regardless of its own expressed views, is mystical. This is true for mechanistics, which cannot deny sensation although it would like to, and which cannot comprehend it, although it ought to. It is of course equally true for any kind of outspoken mysticism and especially for religious spiritualism. But it is also valid for psychophysical parallelism. Therefore, even psychoanalysis, unless it interprets the instinctual drives in terms of concrete, physiologically tangible excitations, is mystical.

Furthermore, the sharp distinction between animism and mysticism results in a sharp distinction in the orientation of research.

The animist proceeds from his own organ sensations, which tell him that organs are motile or alive or, which is the same thing, animated. Although the animist draws direct conclusions from personal experience, he cannot explain anything about the nature of sensation, movement, or animation. Movement is the direct experiential material that shapes the mental images of the newborn child. As long as the child possesses undistorted, naturally functioning organ sensations, he may falsely interpret what is static by animating it. But whenever the naturally perceiving child describes dynamic, living matter, he will judge correctly. If this function is continued later in natural research, it will conclude, as, for instance, Sigmund Freud did, that there is a "psychic energy" anchored in "physical processes." This judgment is correct, for the psychic function is conceived as motion, and motion, in the strictest physical sense, is a shifting of energy. The discovery of cosmic orgone energy proceeded after all from this correct, animistic interpretation. The study of the nature of sensation—purely conceptually and experimentally—led to the discovery of physical orgone energy, which has specific biological functions.

In contradistinction, the mystical concept of the dynamics of emotions can never lead to the discovery of physical energy processes, if only because, in principle, the mystic knows no connection between the physical and the emotional. In practice, mystical man, unlike the animistically thinking and feeling child, does not experience his organ sensations directly but always as if through a distorting mirror. The mystic may be able to describe orgonotic currents and excitations; he may even give details that are astonishingly exact. But he will never be able to comprehend them quantitatively, any more than one can put the mirror image of a block of wood on the scales.

Controlled clinical experience shows that there is always a wall between organ sensation and objective excitation in the mystic's structure. This wall is real. It is the muscular armor of the mystic. Any attempt to bring a mystic into direct contact with his excitation triggers anxiety or even unconsciousness. He can perceive the emotion in himself as in a mirror but not as a reality. This assertion is founded on an experience I had frequent occasions to observe: if orgone therapy succeeds in dissolving the armor in the mystic, the "mystical experiences" disappear. Thus, *the existence of a dividing wall between excitation and sensation is at the root of the mystical experience.*

The mystical experience is seldom found without concomitant brutal, sadistic impulses. Furthermore, to my knowledge, orgastic potency is not found among mystics, any more than mysticism is found among orgastically potent persons.

Mysticism is rooted in a blocking of direct organ sensations and the reappearance of these sensations in the pathological perception of "supernatural powers." This is true for the spiritualist, the schizophrenic, the religious physicist, and for any kind of paranoiac. If a mystical person tries to describe nature with the given preconditions of his character structure, he will only produce a picture of reality that, while reflecting real processes, is not in harmony with objective processes but is distorted: as, e.g., the paranoid schizophrenic's feeling of being influenced by electrical currents, the spiritualist's impression of a blue-gray nebulous ghost, the religious epileptic's fantasy of a "universal spirit," or the metaphysician's idea of the "absolute." Each of these impressions contains a part of the truth: the orgonotic tingling sensations are the "electric currents" of the schizophrenic; the blue color of the orgone is the blue-gray ghost of the spiritualist; the cosmic universality of orgone energy is the "universal spirit" and the "absolute" concept of the mystical character.

Thus, both the animist and the mystic touch upon a reality. The difference is the distortion of reality that becomes the absolute or the grotesque in mystical man, while animation of inanimate matter characterizes the animist. The claims of the mystics are quite transparent and easily refuted. The claims of the animists are hard to refute and more rationally comprehensible. The widespread and acknowledged view of the harmony of nature is basically an animistic view which, in the mystic, is degraded to a personified cosmic spirit or a divine universal being. The mystic is trapped in the absolute. The absolute is incomprehensible. The animist remains flexible, his views can be shifted. He has the advantage that his view of nature, contrary to the mystical view, contains a practicable core of truth. Even now, centuries later, the concepts of the animist Kepler, who formulated the planetary harmonic law, are still valid with respect to his *vis animalis* that moves the planets. THE SAME ENERGY THAT GUIDES THE MOVEMENTS OF ANIMALS AND THE GROWTH OF ALL LIVING SUBSTANCE INDEED ALSO GUIDES THE STARS.

The origin of all animistic and truly religious world philosophies must be sought in the functional identity of organismic and cosmic orgone. Here we also find the rational core of animism and of genuine religiosity; we must liberate this rational core from its mystical guise strictly scientifically, in order to gain the intellectual raw material that leads us to the physical function of cosmic energy. By "physical function" we mean the *orgonomic law of motion,* which must be articulated in orgonometric terms. The poetic and philosophical equation of life sensation and cosmic function is correct but not sufficient to reconcile the human animal with nature. The human animal can learn to understand and love nature inside and outside himself only if he thinks and acts the way nature functions, namely, functionally, and not mechanistically or mystically.

The world of orgonomic "energetic" functionalism is a vigorously functioning, free, and consequently lawful and harmonic world. It has no room for a vacuum in space, which the mechanistic physicist requires because he is inca-

pable of making sense of nature in any other way; neither
has it room for ghosts and phantoms, which mysticism cannot demonstrate. Also, the world of functionalism is not a
"shadow world," as is the world of the abstract mathematician. It is a world that is tangible, full, pulsating, and
simultaneously demonstrable and measurable.

The abstract mathematician does not realize that his formulas can describe objective processes only because his ideas
are part of the same natural function that he expresses as
abstract symbols. Anyone familiar with organ sensations is
capable of tracing the sources from which the "higher"
mathematician, without knowing it, derives his insight. Even
if the functional symbols that he puts in place of the real
world are unreal and do not even pretend to mirror reality,
the creator of these functional symbols is unquestionably a
vigorously pulsating orgonotic system who could not involve
himself with mathematics if he did not pulsate. "Higher"
mathematics could pose as the most sophisticated product in
the development of natural science only because its anchoring
in pulsating nature was not known or not admitted. The
brain of the mathematician is not a differently organized instrument; it differs only insofar as it is capable of expressing
organ sensations in mathematical form. The mathematical
formula is thus only one means of expression among others,
and not the magic wand it appears to be to the narrow mind
of mystical man. It is the living organism that orders, re-
groups, and connects its sensations before articulating them
as mathematical formulas.

The orgone biophysicist knows that in sleep one often
finds solutions to problems one has tried to solve in vain
while awake. I myself have worked out a whole series of
functional equations during twilight sleep, which will have to
be set forth in another context. I do not mind admitting this
because I am not interested in the superiority of "pure intel-

lect" over the "emotions." I know furthermore that the human intellect is only the executive organ of the living plasm investigating and probing the world around us.

Considered functionally, sensation is a feeling out of reality. The slowly groping, wavy movements of animal antennae or tentacles will illustrate what I mean. Sensation is the greatest mystery of natural science. Therefore, functionalism knows its worth and values it highly. Because he regards sensation as a tool, the functionalist is concerned with its care, just as a good carpenter cares for his plane. The functionalist will always order his intellectual activity so that it is in harmony with his "sensations." Where the degree of emotional irrationalism is small—and it must be small for anyone who investigates nature—he listens to the gentle warnings of his sensations that tell him whether his thinking is right or wrong, clear or muddied by personal interests, whether he follows his irrational inclinations or any objective processes. All this has nothing to do with mysticism. It has to do exclusively with keeping our sensory apparatus, the tool of our research, in good condition. This condition is not a "gift," not a special "talent," but a continuous effort, a continuous exercise in self-criticism and self-control. We learn to control our sensory apparatus when we have to treat biopathic patients. Without an invariably clear system of sensation, without the ability to clear it if it becomes irrationally distorted, we would not be able to take one step into the depth of human character structure or describe natural processes as they are.

Such observations and viewpoints in natural research (and man's emotional life is certainly a part of nature!) are alien to the chemist, the physicist of the old school, the astronomer, and the technician. They do not know their sensory apparatus, with which they explore the world. They can control their actions only by experiment, and we know that experiment without organ sensation has taken mechanistic natural science nowhere in crucial questions of nature. The mechanistic technologist denies this, but the eminent physicist admits it.

For us natural scientists, the life function has many meanings:

First, it forms the basis of all life activity, including natural research. It is the port from which we set out on our investigative voyages and to which we return in order to rest, store up results, or pick up new provisions.

Second, the life function is the tool with which we touch, probe, order, and comprehend ourselves and the nature around us. (The German term *begreifen* literally means "to feel out.") The most important tool is sensation, be it inner organ sensation or outer sensory perception.

Third, the life function is an object of our research. The first and most important object is, again, organ sensation both as a tool and as a natural phenomenon. By investigating how living matter functions, we also discover a part of external nature. For what is truly alive in us is itself a part of that external nature. Thus, if we proceed carefully in studying the material that constitutes the life function, we must also find those functions that have general, cosmic validity. This is a necessary and unavoidable conclusion inasmuch as the overall functioning principle is contained even in the smallest special functioning principle.

In this way, the life function becomes for us a part of objective nature, the prototype for certain generally valid natural functions which originally have nothing to do with the living element per se. Offhand, a thundercloud has nothing in common with an amoeba. By observing certain functions in the amoeba, however, we succeed in reaching conclusions that are equally valid for the thundercloud; for instance, there is the attraction of highly charged thunderclouds upon smaller clouds, as compared to the attraction exerted by the amoeba on small bions.

Such rigorously ordered and controlled natural research, such interconnections, are alien and often outrageous to the mechanist. Under no circumstances will he admit any connection between amoeba and cloud, and he will dismiss such ideas as humbug, charlatanism, or mysticism. Therefore, proceeding from the energy development in bions, he does not discover the same energy in the atmosphere. Therefore, as a meteorologist, he describes "heat waves" in the flickering atmosphere as one and the same phenomenon, even at minus 20°C; as an astronomer, he speaks of "bad seeing" and "dispersed light" in observing the stars at night; and, as electrophysicist, he speaks of "static electricity" in dealing with the atmosphere. No one would object, if only he did not believe he had solved all mysteries with the term "ionized cosmic dust." His arrogance turns him into a conceptual roadblock in the field of natural research; and since the further development of the human race will be determined for centuries by its attitude toward nature inside and outside itself, and by nothing else, such ignorant arrogance also becomes a roadblock for any social development. The state of the world today speaks for itself.

Functional thinking does not tolerate any static conditions. For it, all natural processes are in motion, even in the case of rigidified structures and immobile forms. It is precisely this motility and uncertainty in his thinking, this constant flux, which places the observer in contact with the process of nature. The term "in flux" or "flowing" is valid, without qualifications, for the sensory perceptions of the scientist observing nature. That which is alive does not know any static conditions unless it is subjected to immobilization due to armoring. Nature, too, "flows" in every single one of its diverse functions as well as in its totality. Nature, too, does not know any static conditions. Therefore, I believe that Bergson, in his brilliant formulation of the "experience of continuum," made the mistake of describing the biopsychic process as "metaphysics," in contrast to "science and technology." Fundamentally, Bergson meant to say only one thing with his philosophy of nature: mechanistic natural science is correct in the realm of inorganic nature and technological civilization. It leaves us in the lurch if we are to comprehend the perceiving live organism and the act of natural research in the realm of biopsychic processes.

Orgone research has left no doubt that mechanistic natural research has failed not only in the biopsychic realm but also in all other realms of nature where a common denominator of natural processes had to be found. For, as we said, nature is functional in all areas, and not only in those of organic matter. Of course, there are mechanical laws, but the mechanics of nature are in themselves a special variant of functional processes of nature. This remains to be proved.

If we consistently want to follow the working hypothesis that orgone energy is cosmic primordial energy per se; that the three large functional realms, of mechanical energy, dead mass, and living matter, spring from this cosmic primordial energy through complicated processes of differentiation; that, finally, cosmic primordial energy actually functions in a specific varied manner, we face the enormous task of deriving the specific variations from the common functioning principle of orgone energy. We can do this in several ways:

We can study orgone energy in its natural functioning in the atmosphere and in the living organism, grasp the basic functional principles, and trace them to the higher variations. We can simultaneously—indeed, we must simultaneously—comprehend the specific variations in and among the three large functional realms and connect them concretely in such a way that the common functioning principles of the higher order will lead us spontaneously to the common functioning basis of all nature.

This has nothing to do with philosophy or natural philosophy. This task is comparable to that of an engineer who has to build a complicated bridge across a wide river. He must span both banks and construct the bridge in its entirety as well as connect the individual cement blocks. We differ from this engineer in that we cannot promise at what particular time the bridge will be ready. We do not know when the construction will be ended and who will carry it to completion. But in order to keep our perspective, we will have to direct our attention simultaneously to both banks and to those details of the construction that are indispensable for spanning the river. At this point, we do not have to think of the form and the material, the decorative touches, the arrangements for illuminating the bridge, etc.

We have investigated cosmic orgone energy in diverse functioning areas sufficiently to formulate several generally valid principles in the common functioning basis of all nature.

Among these principles, we find PULSATION as the basic characteristic of orgone energy. It can be divided into two antithetical part-functions—*expansion* and *contraction*—or synthesized from them. I realize that I am expressing myself mechanistically. But, for the duration of this study, it is necessary to isolate the function under investigation from the general flow of natural processes, even to let it rigidify, so we can examine it more closely. But under no circumstances should we translate a step, which we had to take because we could not operate otherwise, into an objective property of the function itself. We must not ascribe to nature any properties that are not inherent in it but are seen only at the moment of investigation. It is not pedantry or superfluous warning to say this. Mechanistic natural science is full of such misinterpretations.

The mechanistic bacteriologist stains certain cocci and bacteria with specific biologically effective chemicals in order to make them more visible. Staphylococci react positively to the Gram stain, that is, they appear blue; tubercle bacilli appear red in eosin. The bacteriologist now speaks of the specific color reaction of bacteria as if it were a specific biological property of these microorganisms. This is inaccurate, because the staining is an artificial means of demonstrating the object, and not a specific quality of the microorganism. The cancer researcher of mechanistic orientation has consistently overlooked the true properties of cancer cells because he clings to the secondary, artificial properties that the cancer cells acquired in the process of examination.

The mechanistic physicist says that light consists of seven basic colors, that it is "composed" of them. The functionalist says: *if* I put a ray of light through a prism, it takes on the appearance of a seven-colored scale. Without prism or without a screen formed by rain, i.e., without any artificial interference, light is a unitary phenomenon that has its own specific qualities, such as illuminating a room. And let us not forget that we have not actually understood anything when we say "illuminating."

I can kill an animal and dissect it this way or that. No one would say that the animal consisted of the parts into which I have dissected it. This is especially true in criticizing any kind of mechanistic research. The experimental operation alters the object of research. The coloration of cancerous tissue blots out its living qualities. The dispersion of light through a prism merely indicates how light reacts under the influence of refraction, but not how light reacts without this influence.

In wide areas, mechanistic natural science has fallen victim to the error of thinking that the altered qualities of a natural function are identical with its actual qualities. I explain nothing about the nature of a two-year-old child if I let him make a pattern of triangles and squares. I only say something about the particular situation into which I have placed the child, i.e., how he reacts under this special condition. Things are different if I first observe the child in his natural environment. There the child creates his own conditions of life; he is not reacting to a condition created by me. Therefore, the direct observation of nature is more important than the experiment. To control my observations, I can organize my experiments in such a way that I study *nature*, and *not my modifications of nature*.

I observe that, influenced by concentrated orgone and water, plants will spontaneously grow better than they would in darkness or deprived of water. I act according to the natural conditions of growth if I experiment by irradiating seed with concentrated orgone energy and then compare its growth with seed that has received less radiation or none at all. But if I expose this seed to a chemical solution with which it would never come in contact in nature, I have produced an artificial change of the seed's properties. My result may be useful, pointless, or even harmful. But I have not studied a natural process if I have produced experimental conditions that cannot be found in nature. A child does not by nature place square blocks into the corresponding holes, but plays with sand or earth. By nature, a cancerous cell is not Gramstained but has a natural color of its own. And, by nature, seed functions on the basis of orgonotic processes, and not on the basis of a surplus of potassium.

Every kind of natural observation connects excitation as the cause with sensation as the result, or, conversely, sensation as cause and excitation as the result. That quantitative changes bring about qualitative changes, and vice versa, is a generally accepted fact, just as organic and inorganic life influence, condition, and change each other. Dynamic thinking is therefore no special characteristic of orgonomic functionalism. That natural processes influence cultural processes and that cultural processes change nature is a truism for all thought. By the same token, the interrelated functions of animals and plants, men and machines, males and females, science and art, electricity and mechanics, positive and negative electricity, acids and bases, feudalism and bourgeois existence, mathematics and music, intellect and emotion, thought and experience, etc., are known, recognized, understood, and coped with in practice.

The basic difference between orgonomic functionalism and all other conceptual methods is that orgonomic functionalism not only sees an interrelation of functions but seeks a common, third, and deeper functional relation.

From this logical and simple unification of two functions in a third and common functioning principle, it follows that:

1. In the course of progressive insight, all existing functions become simpler, and not more complicated. Here, orgonomic functionalism is in sharp opposition to all other conceptual methods. For the mechanist and the metaphysician, the complexity of the world increases in direct ratio to the increasing knowledge of facts and functions. For the functionalist, the natural processes become simpler, clearer, and more transparent.

2. With the unification in a common functioning principle, there automatically emerges an orientation of research that presses for knowledge of still simpler and more comprehensive functioning principles. For instance, once we have recognized the common functioning principle in animal and plant, namely the bion, we will encounter, willy-nilly, further and more deeply rooted common factors, such as the common functioning of bions obtained from organic matter as compared to bions obtained from inorganic matter. In this way we acquire a viewpoint from which we can study organic and inorganic nature from one perspective.

It is up to us to decide if we want to examine the special or the general, the diverse or the common, the variation or the basic. The variation has its own functional laws that differ from other variations. At the same time, the variation obeys the general functioning principle of its origin.

In the investigation of the cancer biopathy, the functional

viewpoint gained valuable confirmation. A cancer cell in animal tissue is very different from the amoeba in a grass infusion. Mechanistic research claims that the amoeba stems from germs in the air and that the origin of the cancer cell is unknown. "Amoeba" and "cancer cell" have remained two sharply differentiated areas, both without beginning or end or any transition to other areas. But orgonomic functionalism offered a rich mine of research in comparing cancer cell and amoeba.

The common factors are more important than the differences. Cancer cell and amoeba develop through the natural organization of bions or energy vesicles. The cancer cell is the amoeba of animal tissue, and the amoeba is the cancer cell of plant tissue. Through the interconnection between amoeba and cancer cell, both of which disintegrate into bions in living tissue, a functional relationship is established which opens the previously closed gates to the investigation of the cancer cell and, with it, the disease of cancer. Schematically, this looks as follows:

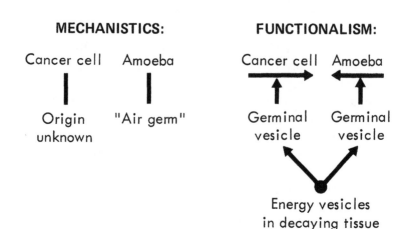

MECHANISTICS:		FUNCTIONALISM:	
Cancer cell	Amoeba	Cancer cell	Amoeba
Origin unknown	"Air germ"	Germinal vesicle	Germinal vesicle
		Energy vesicles in decaying tissue	

Mechanistic thinking favors differences, usually overlooks the common, and therefore becomes rigid and sharply divisive. Functional thinking is principally interested in common features because the investigation of the common leads deeper and further. When Darwin studied the origin of man, derived from the higher animals, he considered it far more important that the embryos of man, pig, ape and dog showed so many identical traits rather than any subtle differences. In this manner he found the common principle of evolution of the vertebrates, which is valid for man and ape. For mechanistics and mysticism, the difference between man and animal was, and still is, more important, e.g., the "non-animal" or "non-sexual" being. We can discern why this divisive methodology was bound to lead to a finalistic and mystical dead end. The common traits are invariably pointing to a common origin. Therefore, the exploration of common functions of different phenomena is also historical and genetic exploration. The divisive observation, as exemplified in purely descriptive biology, cannot lead to genetic observation. Accordingly, the tendency arises to connect the variations with a common "goal" or "purpose" of their functions. This is how mysticism makes its way into natural science. And from the mysticism of divisive observation derive the irrational attitudes of racial prejudice or sexual suppression of infants.

It is no accident, but supported by fact, that life-negating philosophy always emphasizes the divisive element, such as the differences among peoples in nationalism, the differences among families in family ideology, the differences of wealth in the financial principle, the differences of social rank in the authoritarian principle. On the other hand, life-asserting philosophy stresses the common element, the common biological origin of all human animals, the common features in man, animal, and nature, the common life interests and necessities, etc.

Since functional thinking knows the motility of all processes, it is motile itself and always produces an abundance of evolutionary processes. But mechanistic thinking is, by definition, rigid and therefore has a rigidifying effect on the objective of its research, its education, its cure, its social effort. We deny conservatism not its good will but its ability to guide living reality. The mechanist cannot be anything but conservative or reactionary. He may regard his attitudes and intentions in whatever light he pleases; but the essence of his thinking is to overlook developments, to misunderstand or hate the living organism, and therefore to seek a substitute in rigid principles.

The essence of life is to function, and therefore it is antagonistic to any rigidity. Nature knows no bureaucracy. Natural laws are functional, and not mechanistic. Even where the law of mechanistics is valid, nature abounds with variations.

Functionalism is capable of solving contradictions, which seem insolvable to the mechanist, because it comprehends the common principle. To give a few examples:

The mechanist cannot reconcile "society" and "individual," not because he does not want to, but because he is unable to. Therefore, he will give priority to the interests of society or to those of the individual. While he knows that the interests of society are conditioned by gratifying the interests of the individuals, and vice versa, his thinking and acting are invariably a question of *either/or*. This produces the sharp contrast between state and individual, which in this form is insolvable and irreconcilable.

In the sharp juxtaposition of "religion" and "sexuality" we face another example of divisive mechanistic thinking. For the mechanist and the mystic, religion and sexuality are irreconcilable. This is carried to such extremes that, to the Catholic, sexual pleasure is regarded as a sin even in a marriage sanctified by the Church. The functionalist resolves this contradiction as follows. The common principle of sexuality and religion is the sensation of nature in one's own organism. When natural sexual expressions were repressed in the human animal during the development of patriarchy, this produced a severe, unbridgeable contradiction between sexuality as a sin and religion as a liberation from sin. In primitive religion, religion and sexuality were ONE: orgonotic plasma excitation. In patriarchy, orgonity becomes "sin" on the one hand and "God" on the other. The functionalist understands the identity of emotions in sexuality and religion, the origin of the estrangement and the dichotomy it created, the fear of sexuality among religious people, and the pornographic degeneracy among the excommunicated. The mechanist and the mystic are a product of this contradiction, remain trapped in it, and perpetuate it. The functionalist breaks through the barriers of this rigid contradiction by finding the common features in emotion, origin, and nature.

Transgressing the rigid barriers the mechanist has drawn in nature initially takes the functionally thinking scientist to uncertain ground. Mechanistic rigidity in observation and theory formation serves one's personal security far more than it does objective exploration. I have experienced time and again, both in myself and in many of my co-workers, that clinging to rigid barriers and laws has the function of sparing us psychic disquiet. Strangely enough, by letting the motile element rigidify, we feel less threatened than we do in exploring a motile object.

One of my assistants who came to me from a biological laboratory reported that she had received strict orders covering her research work. She was not allowed to go beyond certain limits or step into areas outside the "research program." I realize that such rules rest on the trend of the neurotic character structure to slip into arbitrariness and lack of discipline in thought and work. But I also realize that such rules preclude true research. The bacteriologist, for instance, is so hemmed in by the barriers of sterilization that he forgets that nature is not sterile and that we must also explore processes of putrefaction. We will see in another context that for several decades cancer research overlooked the simple fact of putrescent cancerous organisms because the limits of sterilization were not allowed to be crossed. It is now clear that one becomes uncertain in working with unsterile preparations. But this uncertainty is an essential exercise of balanced thinking. The results produced by "sterile" facts must be compared with "unsterile" facts. This is more difficult but also more constructive. It reduces prejudice and brings us closer to reality.

The exploration of nature by experiment was a decisive step forward toward *objective* observation. But the mechanistically conducted experiment has separated the observer from direct observation. The distrust of man, including his power of judgment and the rationality of his emotions, is so enormous—and rightly so—that the objective experiment became top heavy. One felt averse to both examining live tissues and observing the atmosphere with the naked eye. "Objective experiments," such as Michelson's light experiment, which did away with the ether, have had catastrophic consequences for natural research. It is possible to control the living observer by the experiment, but it is not possible to replace him. An observer, who because of his character structure works and thinks mechanistically, cannot improve his performance by experiments. Hence, it was always the rebel against mechanistics in natural science who transcended the sharp borders and made his discoveries precisely because he was so unorthodox. He simply returned to direct observation and to the natural, i.e., functional, interrelations of these observations. These rebels of natural science were also rebels in thinking; they functioned in an alive manner, stepped across barriers, broke down walls, as in the question of the unchangeability of chemical substances, the relations between energy and mass, the relations between man and animal, etc. Just think of what psychology has accomplished on the basis of these same observations.

The functionalist uses the experiment to confirm his observations and the results of his thinking. He does not replace thought and observation by experimentation. The mechanist does not trust his thinking and observation, and he is right. The functionalist does trust his senses and his thinking. He differs from the mystic and the religious believer by knowing his uncertainties and controlling them experimentally. He differs from the mechanist by including everything in his observation, by regarding everything as possible, by breaking down the barriers between the sciences because he comprehends their interconnections, and by steadily and consistently progressing toward the simpler functioning principle.

The mechanistic scientist is so unsure of himself, his operations are so complicated and entangled in trivial detail having no relation to the whole, that he rejects results a priori as inaccurate merely because they are simple. The orgone accumulator was dismissed by eminent persons because "it is *only* a simple metal box."

The mechanistic human structure has a low tolerance for uncertainties, avoids prolonged tensions caused by uncertainty, does not care for the flowing and intermeshing of functions in nature. Added to this is the fear of life itself, which will be discussed in another context.

By breaking down all barriers erected by the mechanist against nature, by differentiating common functions from specific variations, the functionalist reduces diverse facts to functional interconnections, the functions to energy processes, and the various energy processes to a generally valid functional law of nature. It is unimportant how much he actually accomplishes at any given time in practice or in theory.

What is important is the orientation toward research in the observation of nature. And this orientation (simplification and unity versus complexity) depends on the structure of the scientist.

The mechanistic viewpoint fails when we try to find the transition from orgonotic excitation of the human organism to the processes in the tissues of its organs. The visible spasms and the subjective sensations of current imply that they correspond to concrete processes in the tissue substance. Mechanistics is unable to tell us how we might confirm or control our justified assumption. The processes in human tissues are not immediately observable. Post-mortem dissection and coloration of the tissue do not explain anything about the processes in their living state because dead and dying tissue are fundamentally different from live tissue. The reports of mechanistic pathology are taken from dead tissues that are further changed by staining; therefore, they bypass what is alive and go astray. Also, mechanistics presupposes innervations of tissue functions in man and the higher animals, which are supposed to arise not in the tissues themselves but in the "higher centers." In this way, nothing can be gained by observing primitive plasmatic organisms even if they are accessible to microscopic observation. An amoeba has no nerves and, consequently, no innervations that, from the mechanistic viewpoint, would correspond to those of the higher animals. This is how mechanistic pathology automatically excludes any comparative observation.

Functionalism has freed itself from these prejudices and their rigid limitations. The thought technique of the functionalist connects the animal tissue with the tissue of the protozoon because, in principle, all living substance must be functionally identical. Once the idea of this identity is accepted, there are many experimental possibilities for answering the question: *do the orgonotic sensations, which are so familiar to the psychiatrist trained in orgone therapy, have a real, observable basis in animal tissue?*

Let us observe flowing amoebae. We see currents in the protoplasm which, when pleasurably stimulated, are directed toward the periphery, and which retreat to the center when unpleasurably stimulated. In other words, the amoebae stretch out toward pleasurable stimulation and recoil when unpleasurably stimulated. Here, with one stroke of a simple observation and a sound theory, a solid bridge is built from the multicellular organism to the amoeba. The amoeba behaves exactly as we could have predicted the emotional behavior of the human animal on the basis of our clinical observations. What we discern psychiatrically in man, we observe directly in the amoeba: the flow of protoplasm that has "emotional" significance. Our theory tells us: *what we subjectively perceive and what we call "organ sensations" are objective movements of protoplasm. Organ sensations and plasmatic currents are functionally identical.* With regard to the functions of pleasurable expansion and anxious contraction of protoplasm, man and amoeba are functionally identical.

We let amoebae die off. Their protoplasm gradually loses its motility until it stops altogether. "Death" has occurred. After dying, the protoplasm disintegrates into tiny bodies that we know so well as T-bodies from examining cancerous tissues. The microscopic processes in protozoa have put us on the track of degeneration in the tissues of cancer patients. More than that, if we follow the organization of protozoa from bionously disintegrating grass tissue, we find the key to the origin of cancer cells in disintegrating human tissue. The microscopic observations remain in harmony with our clinical observations. Tissue disintegrates into bions and then into tiny T-bodies when it loses biological energy, i.e., when it becomes anorgonotic. This can be studied under the microscope. These observations correspond to the diminishing life activity in the cancerous organism, the loss of tissue, the typical stale or putrescent odor, the low motility, the resigned character attitude, etc. All this points to steadily progressive orgone loss in the organism. I believe that very few findings of classical medicine rest on such a congruity of diverse facts.

Added to this is the existence of orgone energy in the atmosphere. Concentrated in accumulators, this energy is capable of stopping anorgonotic processes in the sick organism and reversing them. Anorgonia of the blood in cancer patients can be cured by orgone therapy. The organism feels strengthened, it develops stronger impulses, gains weight, etc.

We see that the functional interrelation of facts from different, widely separated areas, achieved by different investigative methods but subordinated to one theoretical principle, is no witchcraft or magic but a technique of thinking that can be learned. Helped by this conceptual technique, we can bridge wide gulfs that up to now have gravely impeded biological and medical research. It is primary biological movement, i.e., the primary emotion, which in a simple manner combines living substance of various organizational strata into *one*. In principle, we have become independent from nerve paths and specific glands because we have put the problem where it belongs: in the foundation of living functioning. Not matter or structure but motion and energy processes are the guidelines of our conceptual technique. Since substances and structural forms are endlessly complicated, while primitive movements and energy processes of life are extremely simple and accessible to observation, we have gained a new and hopeful perspective. At this point, it is the very simplicity of our clinical and experimental perspective that separates us from our colleagues working with chemical substances and structures in mechanistic pathology. Today, simplicity lacks credibility, even if it no longer seems "unscientific," as it did several years ago. I know that the comparison of an amoeba with a man must appear strange to complicated thinking. But I insist that the rigid barrier erected by mechanistic cancer research between the protozoon in grass infusion and the cancer cell in animal tissue strikes me as far stranger.

Scientific research methods prove their accuracy not only by the facts they reveal but also by the new research fields they open up. The mechanistic separation of cancer cell and protozoon has led us nowhere. On the contrary, for decades it has condemned cancer research to sterility. This happened in the name of a prejudice of religious-mystical origin: "the units of living matter are cells, and the cells are forever perpetuated from cells." This prejudice gave rise to the erroneous idea that the cancer cell was merely a degenerate body cell. *The cancer cell has nothing in common with the healthy cell, except that it develops from the decayed matter of formerly healthy cells.*

In contradistinction, the functional connection between the cancer cell and the protozoon in decaying grass tissue has opened the gates to further cancer research.

With this basic attitude and thought technique we are spared fruitless discussion about the biochemical results of classical biology. They are of secondary importance for understanding living matter and, with it, the cancer biopathy. An example from the field of mechanistics, which is more familiar to the mechanistic thinker, may illustrate what we mean.

A railroad train consists of a number of cars drawn by a locomotive. The cars are made of metal, wood, glass, etc. The locomotive houses a firebox, a steam boiler, levers, pistons, etc. No matter how much we say about wood, metal, glass, levers, etc., no matter how closely we analyze them in detail, the most exact investigations, carried on ad infinitum, would still not tell us anything about the function of a railroad train. Its one and only function is to move *in toto* and to take me from New York to Boston. If I want to understand the railroad train, I must understand the principle of its motion. The material construction of its locomotive and cars is unimportant and of secondary interest; it is perhaps of interest for the comfort and safety of the journey, but not for the principle of traveling.

Now classical biology examines the structures of living matter in its infinite variations down to the smallest detail. It may produce results of great sophistication, but it will never be able to say anything about the nature of living matter.

We are dealing with more than questions of biology. The discovery of the orgone far transcends the realm of living matter, even if it stemmed from this realm and found its most important application therein. As we said, the discovery of the orgone must essentially be attributed to a complicated but consistent thought technique. This thought technique was confirmed by the findings it made, and by the development of experiments that secured the orgonomic findings. The description of this act of thought becomes an integral part of understanding life itself. In it, life understands its own essence.

I say: *In the act of thought, life comprehends its own essence.* This is true for the functions of both inorganic and organic nature. In building a machine, man grasps the laws and functions of non-living nature in its relation to living requirements. In the sciences concerning man, the living organism seeks to understand the functions of life itself. However, it always understands only what it experiences in itself. If what is alive in the human animal had not become armored and degenerate because of the mechanical-mystical principle, the result of mastering living nature would be in harmony with actual life functions. It would have mastered the material structures of living substance side by side with the laws of motion of the living organism. Owing to the social tragedy that struck the human animal thousands of years ago in the form of mechanical-mystical degeneration, it had access only to its mechanistic functions, to the structure of the skeleton, to the muscles, the blood vessels and nerves, the chemical composition of the organism, etc. Since the motile aliveness in man was armored and therefore inaccessible, the life principle itself, the motion, i.e., actually the most essential feature of life, remained a closed book. What the rigid mechanist could not accomplish because he regarded life merely as an especially complicated machine, the mystic has tried to supplement; the motility of life was transferred into the beyond, allegorically in theory and so often literally in practice whenever rigidified human animals went to war against one another.

Because armored man is rigidified, he thinks predominantly in terms of matter. He perceives motion as being in the beyond or as supernatural. This must be taken literally. Language always expresses the immediate condition of organ sensations and offers an excellent clue to man's self-awareness. Movement, i.e., plasmatic current, is indeed inaccessible to the rigidified human animal. It is therefore "beyond," i.e., beyond his ego perceptions; or "supernatural," i.e., felt as an eternal cosmic longing *beyond* his material being. What the armored organism perceives as "mind" or "soul" is the motility of life that is closed to him. He sees and feels the motion only as in a mirror. He describes the motility of life correctly, but only in the sense of a correct mirror image. A large part of the brutality of the mystic is explained by the simple fact that while he feels life inside himself, he can neither experience it in reality nor develop it. Hence the impulse develops to conquer the mirror image by force, to make it tangible and palpable by force. The life in the mirror is a constant provocation that drives him into a frenzy. There it is, this motility; it lives, laughs, cries, hates, loves—but always only in a mirror. In reality it is as barred to the ego as the fruits were out of reach for Tantalus. From this tragic situation springs every murderous impulse directed against life.

Mechanistics and mysticism combine to form a sharply divided image of life, with a body consisting of chemical substances *here,* and a mind or soul *there,* mysterious and unexplorable, inaccessible as only God himself.

The unarmored organism, however, experiences the self mainly as a unity in motion. Its organ sensations tell it that the essential part of life is not substance. Basically and in terms of matter, a corpse looks no different than a living body. Until putrefaction sets in, the chemical composition is the same. The difference lies in the absence of motion. Therefore, the corpse is alien, even terrifying, to living sensation. *Spontaneous motion is thus what is alive.* We now understand the hopelessness of all mechanistic-mystical thinking. It constantly collides with the armor of its own organism without ever being able to penetrate it.

Unarmored life, however, will find, interpret and comprehend expressions of life in its own movements. Motion is its essence; structure is important, but not basic. Therefore, the biology of the unarmored organism must necessarily differ radically from the biology of armored life.

The mechanist does not understand the principle of human organization. He does not know the properties of orgone energy, and therefore, unless he remains purely descriptive, he is forced to introduce a metaphysical principle. For him, there is a hierarchy of organs in the body. The brain as the "highest" product of development, together with the nervous system in the spine, "directs" the whole organism. Mechanistics postulates a center from which all impulses proceed to set the organs in motion. Communicated through the corresponding nerve, every muscle has its own center somewhere in the brain or midbrain. How the brain itself receives its assignments remains a riddle. The organs are the well-behaved subordinates of the brain. The nerves are the telegraph wires. Hence the coordinated movement of the organism remains nebulous and mysterious. Where understanding fails, "purpose" sets in, the convenient "in order to." The muscles of the shoulders and arms of the apes coordinate their movements "in order to" grasp. To my knowledge, a "center" for the coordination has not been assumed or found. And it would not improve the situation, because the question of who gives the assignments to this center would still be unresolved.

Since the mechanist does not understand the living organism, he must resort to mysticism. Therefore, all mechanistic philosophy is, and invariably must be, mystical as well. Mechanistic thinking itself is clearly made in the structural image of social patriarchy when it regards the brain as the master, the nerves as the telegraph wires, and the organs as obedient executive subjects. And behind the brain there is "God," or "reason" or "purpose." The situation in the scientific comprehension of nature remains as hopelessly confused as ever.

In functionalism, there is no "higher" center and no "lower" executive organ. The nerve cells do not produce the impulses; they merely communicate them. The organism as a whole forms a natural cooperative of equivalent organs with different functions. If natural work democracy is *biologically* founded, we find it modeled after the harmonious cooperation among the organs. Multiplicity and variety are fused into unity. *Function itself regulates cooperation.* Every organ lives for itself, functions in its own realm on the basis of its own functions and stimuli. The hand grasps and the gland secretes. *The individual organs are independent beings endowed with their own sensation and function.* Experiments with the isolated heart and muscle have unequivocally confirmed this. Sensation is by no means tied to sensory nerve endings. All plasmatic matter perceives, with or without sensory nerves. The amoeba has no sensory or motor nerves, yet it perceives.

Each organ has its own mode of expression, its own specific language, so to speak. Each organ responds to stimulation in its own specific way: the heart with change in heartbeat, the gland with secretion, the eye with visual impressions and the ear with sound impressions. The specific expressive language of an organ belongs to the organ and is not a function of any "center in the nervous system."

In confronting these two basic views of the organism, we clearly recognize the difference between unarmored and armored living matter. Each derives its judgments from the organ sensations of its own body. The unarmored organism grasps directly with its hands. The concert pianist does not give orders to his hands. The hands, in conjunction with the whole organism, are the moving and movable self-acting organs. One hears with the whole organism, not just with the ear. The wheel is not the automobile. One travels by car, not by wheel.

The armored organism, on the other hand, perceives the self as consisting of isolated parts. Every impulse must penetrate the armor. From this, the feeling of "you should" or "you must" arises, as well as the idea that the organism has a higher center that gives "orders" to the executive organs. In addition, there is the sensation of heaviness, inertia, or even paralysis in the limbs and the torso, which gives credence to the idea that an organ must act and be moved by an order. By the same logic, there is an "ego" behind all this, an intellect, a "higher reason," which "guides," "assigns," etc. From here to the political concept of human society or, conversely, from the concept of the absolute state to the mechanistic concept of the organism is only one step.

This is the way the armored organism developed and still develops its concept of living processes. Furthermore, the divisive idea of its organs and sensations makes it unable to find functional connections, which explains how brain mythology dominated the natural sciences for decades without anyone ever realizing that billions of organisms functioned for untold millennia before there ever was a brain. In addition to the splitting up of organ sensations, there is the mortal terror of total pulsation, of spontaneous motion and spontaneous excitation. This anxiety constitutes the actual brake. If the splitting-up process prevents the functional unity of the individual functions, the anxiety produces terror or rage in the armored organism whenever another person fills and connects the gaps, comprehends functional unity, or creates it.

For these reasons, classical biology did not progress beyond the cell and did not find the simple path toward proving that cells are organized from bions and decompose into bions at death. The armored organism is characterized mainly by its inability to perceive and feel motion, i.e., living functioning, and therefore to comprehend it. What is usually called the rigidity or conservatism of traditional science is in reality marked by this inability and fear on the part of prominent scientists, who are then imitated by a multitude of minor ones. We know many examples of such dogmas: the indestructibility of the atoms, the division of matter and energy, *omnis cellula ex cellula,* etc. A great deal has been written on these subjects. But here, for the first time, these dogmas are successfully understood and, thereby, shaken. Many other dogmas will be destroyed in the further development of functional thinking.

Unquestionably, the most important distinction between armored and unarmored orgonotic systems is the development of destructive sadism in the former. Since, in the armored organism, every plasmatic current and orgonotic excitation, in reaching for contact, runs into a wall, an irrepressible urge develops to break through the wall no

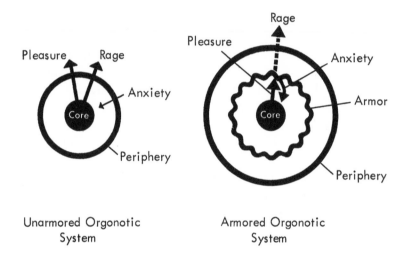

Unarmored Orgonotic
System

Armored Orgonotic
System

matter by what circumstances or means. In so doing, all life impulses are converted into destructive rage. The organism tries to break through the armor by force, as if it were imprisoned.

I seriously believe that in the rigid, chronic armoring of the human animal we have found the answer to the question of his enormous destructive hatred and his mechanistic-mystical thinking. We have discovered the realm of the DEVIL.

CHAPTER V

THE REALM OF THE DEVIL

All true religion corresponds to the cosmic, "oceanic" experience of man. All true religion contains the experience of a unity with an omnipresent power, and simultaneously of temporary, painful separation from this power. The eternal longing for return to one's origin ("return to the womb"; "return to the good earth from whence one came"; "return to the arms of God," etc.), for being embraced again by "the eternal," pervades all human longing. It is at the roots of man's great intellectual and artistic creations; it is the core of his longing during adolescence; it pervades all great goals of social organization. It appears as if man yearns to comprehend his separation from the cosmic ocean; such ideas as "sin" have their origin in an attempt to explain the separation. There must be a reason for *not* being united with "God"; there must be a way to unite again, to return, to come home. In this struggle between the cosmic origin and the individual existence of man, the idea of the "devil" somehow arose. It is the same whether one calls it "inferno" or "hell" or "hades."

The human race knew nothing about the "ether" in the physical sense. It perceived the ether as "God," "prana," "entelechy," etc. Its ideas of a "better future" or of "paradise" centered on the idea of "unity with God." But God, the representative of the living process to the mind of man, could not be reached and remained forever inaccessible. Being only a tiny fraction of the cosmic orgone ocean, the animal, man, could not hope to reach the all-inclusive creator. What he could do, however, was to hope for salvation through the resurrection of the messiah who would free him from his sins and unite his soul again with the cosmic orgone ocean ("nirvana," "return to God"). Why, we must ask, did man not feel united with his God to begin with? Why did he feel sinful? Why was redemption necessary as in the Christian religion or severe punishment as in the Jewish religion?

What kind of reality underlies the idea of the "devil" and similar fantastic products of the human mind?

Most religious philosophies and scientific investigations of religion failed to explain the devil because they moved inside and never outside the realm of prevalent human thinking. They did not describe or investigate the human character from the standpoint of its cosmic environment, but, on the contrary, they explained the cosmic in terms of human nature. Accordingly, there existed such an entity as the devil or some other type of malicious being as the counterpart of God. God was the *good* and the devil was the *evil*. God was unreachable, unknowable, beyond the grasp of the human mind; but man's soul was caught in the devil's clutches. God and devil were absolute opposites. Both ideas resulted from great errors. Both originated within the human character structure. Both were very powerful obstacles in man's way to understanding his own true nature, and to realizing himself.

These simple functions of life remained inaccessible:

The natural work relationships between men. They actually were the foundation of man's existence, but he did not know it. Moreover, he did not understand and thought it peculiar when he was told about them. On the other hand, that which had no foundation in man's existence, the politician, the knight, the king, the representative of God against the devil, etc., occupied man's mind through the ages.

The working man carried a heavy responsibility when he built bridges or drove trains or educated children. He actually had this responsibility in every movement of his body, but he did not know it. He thought he was a nobody and that only his boss, the judge or the chief of police were the ones burdened with social responsibility.

The working man saw nature functioning and growing in his children through the millennia. He saw his children being born as little animals with genitals and natural longings. But he did not understand this and he punished his children for being animals. True, God had created everything, including the genitals. It was obviously because of genital functioning that children were born at all. But, on the other hand, to have genitals was shameful, somehow due to a devilish institution; to touch them was a great sin. For thousands of years a tremendously powerful organization preached that pleasure in the genital union was sinful. And man believed it, did not feel his own body, did not trust his own senses, neglected his very origin and lost the key to his fecundity.

The working man held in his hands all the power he needed to be truly free, but he did not know it and he gave it away to a master. He could have prevented every single war in the history of mankind, but he did not know that he could do it. His real life was here and his ideas about his life were there. What was making life run its course was despised: manual labor, adolescent love, children's genital games, the joy of life. What was set up to kill life was highly honored: the emperor, the Jesuit, the professional killer of people. When he had gained some maturity in political matters, man voted for a secretary, but he did not vote for or against war. Man was full of sexual curiosity and misery; his social parties and newsstands and dreams were bursting with "sex."

But he had banned from his universities the knowledge of the orgastic plasma convulsion and of life.

What does all this mean? What sense is there in this non-sense? There must be some sense in it as in every type of irrational behavior. One cannot improve man's existence by blaming this one or that one for this or for that. "Passing the buck," as American slang puts it, won't help. I am at odds with my friends who correctly criticize the status quo in social affairs but do not go back to its common denominator, the armored, biologically disrupted human structure that produces this status quo. At the risk of appearing "one-sided" or "fanatically aggressive," I venture the statement that *most philosophies about human misery are built on the evasion of the essential.* I venture, furthermore, to assert that every single human being living and yet to be born knows or will know exactly from where all the misery stems. But just as an American farmer thinks he is a nobody and that a muddle-headed mystic of a past vice-president is somebody, every single human being knows the truth about himself and his world but deems himself unimportant.

The realm of the devil is a vicious circle. The harder you strive to get out of it, the more you become stuck in it. This is not a *bon mot* or a joke. It is deadly serious. The devil is an essential function of the armored animal, man. Therefore, let us again survey his basic characteristics.

Armored man is shut off from immediate contact with nature, people and processes. Therefore, he develops a substitute contact, which is characterized basically by lack of genuineness. The larger a city, the more lonesome the individual in it.

Every love impulse meets the barrier of the armor. In order to express itself, it has to push through the rigid wall by force; thus it is inevitably transformed into cruelty and hate.

The original love impulse will, in conjunction with the later hate impulse, appear only as a general attitude of hesitancy, ambivalence, self-loathing, and dependency on everything that promises redemption or release of tension.

The body armor makes inaccessible the basic organ sensations, and with them the genuine feeling of well-being. The feeling of one's own body is lost and natural self-confidence with it. They are regularly replaced by fake, show-off appearances and false pride.

The loss of natural self-perception splits the person wide open into two opposite and contradictory entities: The body *here* is incompatible with the soul or the spirit *there*. The "brain function," the "intellect," is split off from the rest of the organism; the latter is "mastered" as the "emotional" and the "irrational." What is sad about all this is that *within* the framework of armored man's existence, everything is correct and logical.

Since a layer of viciousness is interpolated between the deep, natural core ("God," "Jesus," "the good," "one's soul," etc.) and the superficial appearance, the original "goodness" is shut off and becomes inaccessible. Therefore, quite logically and correctly, the emotions are considered "bad" and the intellect is "good." Co-existence and cooperation of sane emotions and a sane intellect are unthinkable. All institutions of the armored human animal are geared to this dichotomy. The living function becomes perverted into the *mystical,* and the "brain stuff" into the *mechanical* kind of existence. The "bad" instincts are kept in check by "good" morals. Again, this is perfectly logical and correct *within* the given framework of thinking. Those who only curse the moralistic structure of our society without seeing and understanding the logic in it would fail miserably if they were to take over the rule of society and the human masses. The bad instincts are summed up under the heading DEVIL, the moral demands under the heading GOD. Thus, God is fighting the devil, and the devil is eternally tempting poor man to sin against God.

Apart from the mass of diseases it creates, the process of armoring in early childhood makes every living expression edgy, mechanical, rigid, incapable of change and adaptation to living functions and processes. The organ sensations, which have become inaccessible to self-perception, will, from now on, form the basis of the total realm of ideas that center on the "supernatural." This, too, is tragically logical. Life is beyond reach, "transcendental." Thus, it becomes the center of religious longing for the savior, the redeemer, the "beyond." Just as the organ sensations have become inaccessible, the intellectual ability to grasp what is alive also has been blocked. Furthermore, since the sealed-off realm of life manifests itself in the form of anxiety whenever self-perception tries to break down the rigid barrier, the longing for the "beyond" soon acquires two allies: one is *brutality,* which stems from the continuous effort to break through the rigidity of the organism; the second is deep *terror,* experienced as fear of extinction whenever one is reminded of the "lost paradise." Therefore, it is again quite logical that the armored *Homo normalis combines into one mysticism, brutality* and *fear of the natural life functions,* especially the function of the orgasm. The ideas of the "absolute," the "eternal," the "sin" also result from this split of the personality as discussed previously. The "absolute" mirrors the rigidity; the idea of the "beyond" mirrors the inaccessibility of the biological core; the brutality is an expression of the continuous attempt to break through; and the deep-seated fear of the living tells us that armored man has become incapable of functioning in the natural self-regulatory way of the genital character.

We can pursue the social anchoring of this split through the entire written history of mankind, through its religions, its morals, its eternal wavering between law and crime, between absolute authority and irresponsibility of the working masses of people.

Whoever realizes that our civilization (and any similar civilization, for that matter), which developed from this structure of man and his society, is disintegrating will not hesitate to agree that no ideology of guilt or morals will ever solve the tragic contradiction in man's existence. One cannot break the vicious circle by enforcing either one of its two components. If one tries to increase the morals, the perversions and the brutality will increase. If one uses brutality to overthrow the morals, the result will be more and stricter morals, as in imperialist Russia in the twentieth century.

In order to break through the vicious circle in which man finds himself entangled, it is necessary

1) *To understand and acknowledge the rational and useful as well as the useless in the irrational* within (and *only within*) the given framework of living and thinking. Without such an understanding, every attempt to improve the human lot is bound to fail and to end in worse conditions than those it was to abolish.

2) *To stop proclaiming new programs and new political platforms.* Man has run away from himself ever since he began to be aware of his misery and to long for freedom; therefore, one program after another was proclaimed, and one after another failed miserably. The fault is not with the programs but with the running away from their true fulfillment. Every major religious or social movement was initially rational. Every one of them failed and degenerated sooner or later, developing more or less cruelty in the process. Each new movement blamed someone else for the misery of man. The Christians blamed the Jews and the Jews blamed the Christians. The bourgeois blamed the feudal caste and the feudal caste blamed the plebes. The proletariat blamed the bourgeois and the bourgeois blamed the proletariat. It is

high time to stop blaming. It is time to search for the common denominator of all this holocaust of messy thinking. It is time to go back to the origin of all the excellent human attempts to better man's condition. One will find that these many attempts in ideas or programs or political ideologies are not so far apart as is believed by those who think that everyone is to blame except themselves.

The common denominator of all these cruel failures is man himself, who cut himself loose from his own nature. Whatever he takes over is bound to perish as long as he does not finally attack his own biophysical structure. *And this is no longer a question of "politics" but of the disarmoring of the human animal, of how our newborn babies grow up.*

In the early 1920's, inchoate sex-economic research had unequivocally demonstrated the natural functions in the small child that govern and safeguard self-regulation. *The core is the biosexual function centering on genital development.* Self-government in the realm of social processes is entirely and basically dependent on the natural self-regulation in each newborn infant. Man ran away from this simple, clear and decisive fact again and again—into the artificial settings of experimental psychology, into the "cultural adaptation" of psychoanalysis, into the class fraud of the social mass movement.

Alexander Neill has for decades done a good job in proving practically my contention that natural, self-regulatory development of children *is* possible. Bronislaw Malinowski has corroborated the same contention through his studies of Trobriand society. The function of self-regulation is no longer the problem. The main problem now is—and will remain for a long time—how to safeguard this natural growth of children, how to protect it against a type of public opinion that stems from the armored, rigid, lifeless, fearful, hopeless animal, *Homo normalis.*

It is clear that man must stop running away from himself,

from his own programs, platforms, intentions, and abilities. People talk too much, write too much, quibble too much in order to outscream their own inner emptiness, in order to evade the main issue: the *reason for the big evasion, the runaway.*

3) *To step outside the framework of thinking of the "absolute,"* the eternal values, the antithesis of God and devil, good and bad, intellect and emotion, to step outside, take a breath of fresh air, and think things over again in a basic fashion. In this basic, painful process of reorientation, it is necessary to behave as a skilled surgeon or a good psychiatrist behaves—to keep free of the irrational maze and chaos of the world of *Homo normalis*. Once one has established one's standpoint outside the holocaust, things become simple: what was oppressive before appears as sick, what appeared as sane becomes insane, e.g., the idea of premarital chastity or sexually innocent children or the nightmare of children's misery.

Standing outside the given social and moral set-up does not mean anarchy or a merely negative aloofness. On the contrary, it means looking at things that happen, at ideologies, party programs, platforms from the viewpoint of unarmored life. From such a viewpoint, the senator who sings a song before a convention to show off how "popular" and "democratic" he is in order to capture more votes looks perfectly ridiculous. The same is true for the mystic who discovered Russia not when it struggled hard for a new life in 1918, but in 1948, after a cruel, vicious, sly dictator—let's call him "Ivan the Horrible"—had led two hundred million people into utter desolation. The problem is how to protect our efforts to make a new life for our newborn babies against the stupidity of any type of "candidate" who irresponsibly promises paradise on earth without burdening the working people with all the responsibility for their existence that they can carry, and against the stupidity of the thousands who hail him. The problem is how to step over the borderlines of present-day thinking and how to keep safely enough outside to survey the territory and to look for ways and means of adjusting man and his society to the principles of life, and not the principles of life to the state idea of a nuisance-making, criminal, good-for-nothing politician. The problem is how to guard this great undertaking against old, frustrated spinsters, against neurotic cranks in high positions, against red fascists, against self-appointed government officials who "investigate" what is none of their business, concerning which they know nothing, thus causing months of anguished disruption of life-important work, crucial for millions. The problem is what kind of organizational effort is needed to bring the members of the working population back to themselves, to free them from the innumerable bad, disgusting, ruinous goings-on in the world of politics, dishonest business, neurotic education, cowardly medicine; how to help them learn to govern themselves without falling prey to new dictators, new horrible politicians, new cranks, or ideologists. These are the problems and there are many more of them. Thus the problem is not what could and should be done; that would be comparatively easy. The real problem is how to start the doing.

As obstacles in our way, we shall find the surgeon who by mere chance became the head of a psychiatric institution and who administers shock therapy ("Shock him!" he says, when a sex-starved schizophrenic attacks a nurse, instead of understanding the basic reason for such attacks); the religious fascist who thrives on human guilt feelings and misuses them to suppress still harder and more cruelly, who does not know a thing about life or love or children or work or accomplishment; the neurotic, snooping, inspecting state official who says he is a paid servant of the working people only when he campaigns for votes; the professional authority who never touched a "hot potato" like adolescent love life during his whole existence but has a set, though disastrous, opinion about it. *And, finally, we shall find as an obstacle in our way the fear of life and love and the majestic simplicity of natural functioning on the part of the people themselves,* the mothers and fathers and teachers and doctors and nurses. We shall see and experience how easy it is to set up new programs and how difficult it is to handle a single case of an adolescent girl who becomes pregnant because of criminal neglect on the part of some Department of Education.

This process of finding the hard way to the simple answer; this practical stepping outside, staying outside and not perishing outside on the part of entire groups of educators, physicians, social workers, parents, adolescents, and CHILDREN will be the greatest and the most radical revolution in our lives or in the lives of our children and their children. This, and not the idiocy of an Ivan the Horrible who plays a game with human lives in some "corridor" over some big city. What an age in which to witness such monstrosities and stupidities; to witness them and not to be able to get them out of the domain of this planet into some Tenth Reich of their own creation. . . .

. . . . Not to be able to get them out . . . , this is the aching worry, the greatest of all social problems. The world of the human animal is full of burning practical issues that await solution, of justified expectations that can and one day will be fulfilled, of longings that would carry this world of ours far beyond present-day imagination of even the highest order. But so-called governments and educational bodies are busy with petty formalities. There must be a common basis of the inferno. Man has become incapable of reaching what he most sincerely desires. God, the "good," peace, cooperation, international brotherhood, happiness are unreachable goals. The devil rules the world. Can biopsychiatry give the answer? I think it can. But here again the devil will prevent this answer from becoming living reality. The right answers

were given many times by many men through the millennia.
The fault was not with the answers but with the devil. The
answers are clear and simple, ready for practical application.
The problem is the obstacle in the way, the devil and his
realm, the inertness of the armored human animal and the
present impossibility of penetrating its armor and making
it think and act rationally.

What now do we mean when we say "devil" as opposed to
"God"? When I say "devil" I mean exactly the same thing
the Christian or the mystic speaks of when he describes
"evil." The core of the matter is the deep anxiety in the organism, so-called orgasm anxiety, which keeps man from realizing himself and his aspirations. We know it is the armoring of the human animal that threw it off the path of
rational biosocial living. But we do not understand yet why
man did not realize this long ago and why he runs berserk
whenever his armoring is challenged. WHAT CONSTITUTES
THE OBSTACLE IN THE WAY TO RATIONAL LIVING? WHAT IS
IT THAT MAKES THE PROCESS OF DISARMORING SO DIFFICULT
AND SO DANGEROUS?

It cannot be the armor itself. People generally understand
what is meant when we speak of "characterological rigidity"
and "muscular armoring." They understand and appreciate
this piece of knowledge better than any other psychiatric
teaching. That my work has survived several disasters and
many threats of destruction is due primarily to this sympathy
and understanding on the part of the armored human being.
Therefore the armor itself cannot be the main obstacle to
our efforts. What then is it?

Let us seek the answer in our medical office. Let us survey
the most illuminating, most impressive experiences with the
armored human being in order to find the answer to this all-
important question.

As we survey our patients, students, co-workers, people in
their usual environments, and as we ponder their most com-

mon and typical reactions, reactions they have in common
with generally disastrous human behavior, we are struck by
nothing else so much as by the TERROR that seizes the
armored individual when he comes into contact with his
biological core, with what we call the plasmatic currents.
Orgone biophysics has termed this reaction "orgasm anxiety." We would be understating the situation if we assumed
that orgasm anxiety is only one among other anxieties of
man, that it is only one of the many peculiar, life-inimical
reactions that have become so well known to modern biopsychiatry over the past twenty-five years. Orgasm anxiety is
much more than, and very different from, say, a simple
phobia or a neurotic anxiety attack. Orgasm anxiety is to a
simple neurotic anxiety reaction as a flood that inundates
millions of acres of farmland and takes thousands of human
lives is to a break in a water pipe in our home.

A simple phobia is limited to a single object or a single
situation, to a knife or to a dark room. Orgasm anxiety is an
over-all biological experience from which there is no retreat.
A simple phobia may annoy people and limit the activities
of the sick individual. Orgasm anxiety goes together with
the experience of a total loss of one's personality and orientation in life. Suicide is rare in simple anxiety attacks if the
main wall of the armoring remains intact. On the other
hand, orgasm anxiety is regularly accompanied with the danger of a complete breakdown, which can become so unbearable that there is no other way out than suicide. Many a
suicide is due to such a sudden, overwhelming breakdown of
the protective armoring. If the individual is being treated by
a well-trained orgone therapist, who knows well the symp-
toms and the processes involved, and who handles them
carefully in single, cautious steps, the danger is reduced to a
minimum; it makes the situation safe enough. But when an
individual whose armor is breaking down in an over-all fashion is left to himself, suicide, murder, or psychotic break-

down is the most likely outcome. It is the sudden loss of control over the deep forces of the biosystem that constitutes the danger. It is furthermore, or rather first of all, the incapacity of the organism to deal with the full force of natural bio-energy that makes the situation in such cases so dangerous. The individual who from childhood has been accustomed to intense emotions and who does not possess strong secondary drives is not in danger when strong emotions develop. But the individual who was armored all his life and never felt emotions strongly or who had only the outlet of neurotic, symptomatic energy release lapses into utter disorientation and despair when he has suddenly to face the full vigor of his bio-energy. In addition, the healthy individual whose bio-energy is discharged regularly in the genital embrace never develops the amount of energy stasis that would add the surging energy impact of pent-up emotions to the danger of the breakdown of the armor.

To summarize: The incapacity of the armored biosystem to cope at all with strong bio-energy, the great amount of dammed-up energy due to life-long stasis, and the quite different nature of deep biophysical functioning as compared with the superficial everyday living of the armored individual constitute the danger. Thus, the armor has a very important function to fulfill, as pathological as this function actually is. It renders protection against a situation that, though natural to the unarmored human, amounts to nothing less than disorientation in the chronically armored human. What we call "freedom giddiness" results from this inability of the armored organism to function naturally. We see "freedom giddiness" in children as well as in adults who are too suddenly transplanted from an environment that functions entirely in accordance with principles of armoring into an environment relying on natural principles of self-regulation. If today or tomorrow the authoritarian state organization were suddenly abolished so that people could do as they pleased, chaos, not freedom, would result. Years of utter disorientation would have to pass before the human race would learn to live according to the principles of natural self-regulation.

This deep-rooted biopathic organization of man appears to the careful student of human behavior as the most outstanding reason for the failure of all previous attempts to secure human freedom. This fact is unknown to the politician who promises freedom and heaven on earth in a quite irresponsible manner; he is the first to run if what he promises really happens. It is quite alien to the average educator, progressive or otherwise. It is the real reason why such methods of bringing up children as employed by Alexander Neill cannot be used by other educators and therefore are restricted to self-governing islands like Summerhill School. Only the right person can do the right thing and only the wrong person does the wrong thing. Only human beings with a structure capable of freedom can live in a self-regulatory, truly free manner. Transplant a child who was brought up by armored human beings into a free atmosphere and one will soon be convinced that *social self-government requires self-regulatory character structures*.

I emphasize these facts not because I am against freedom, but, on the contrary, because I am all for it. If I wish to build a house on a certain piece of ground I must know on what kind of ground I am building. Are there rocks in the depths or is there only mud? If I know there is mud underneath, I still can drain the region and lay in rocks. But if I were irresponsible enough not to know the foundation of my building, I would fail completely. In the realm of education the situation is especially dangerous, since it gives the authoritarian educator a good excuse to continue with his methods of squeeze and drill. He would rightly claim that my feedom method simply does not work, and I could not provide proof to the contrary. One does not remedy the situation we are all in if one closes one's eyes to the depth and scope of the emotional plague that has reigned over social living for millennia. We are all for liberalism and liberals. But we must regret their reluctance to face squarely the issue of human degeneration. Public opinion is sharply divided between those who claim that man is "completely good" and those who claim that he is "bad through and through." I believe that our well-founded knowledge of human structure comes close enough to the truth to make possible changes for the better. We do not give up all hope as does the authoritarian who lacks knowledge of the deep, self-regulatory functions; and we do not become easily disappointed when we look at the almost equally deep-seated inability of man to live freely.

To return to our main line of thought:

It is not the armoring itself that keeps the human animal from reaching its goals of freedom, happiness, and prosperity; the human animal would have learned long ago how to eliminate its armoring had it been the sole cause of the suffering. No, it is the utter disorientation and the threatening breakdown of its whole being, social as well as individual; it is the terror of facing a totally different kind of living; it is the layer of cruelty and hate that lies between man and his goal of peace and goodness; it is the complete anxiety-ridden loss of his biological life-orientation that constitute the great obstacle. Religion and mysticism were not wrong in claiming the existence of the realm of the devil; they were wrong in not looking *beyond* the devil, in not recognizing that God, the eternal, unknowable, unattainable God, is a reality that has been perverted into the devil. Dante's inferno is unsurpassed in its description of the realm of the devil. But even this great work was stuck within the boundaries of a few thousand years instead of going beyond man himself and finding his roots in the wide realm of his natural origins. We must not wonder at the great error committed here. The realm of the devil is so horrible, the depth of human structure so filled with thoroughly antisocial and criminal impulses that whoever dealt with this realm thought it to be the last and deepest possible layer of human life.

Orgonomy succeeded in going beyond the realm of the devil, not because of a special inspiration or supernatural sense but solely because of the faithful and conscientious study of the function of the orgasm. This function has its roots in cosmic, orgonomic laws and, therefore, not only governs the whole of the living realm far beyond man, but in addition, represents exactly what the truly religious man calls his "unreachable God." Orgonomy succeeded in going beyond the "devil" because it learned to master the terrifying obstacles that are piled up in the way of every individual who transcends the realm of the devil, i.e., the realm of the unconscious secondary drives. Once one has reached rock-bottom in the natural orgonomic function, which is represented in the biosystem as the orgastic convulsion; once one has mastered the sharp distinction between the deep biophysical functioning and the distortion of life in the realm of the armoring, the devil begins to lose most of its horrible aspects. We then look at the devil "from below" as well as from beyond and not "from above," e.g., from the standpoint of a national or ecclesiastical interest. In the language of the true Christian one would say, without becoming mystical, that the "devil" is driven out by the functions of "God" or "Jesus."

I am intentionally expressing myself in this manner in order to convince the reader that there are great truths in such religious teachings, even if they have been distorted by the armored human animal.

The "devil" meant the absolute "evil," personified in the well-known creation of hell in Christian thinking and so splendidly embodied in the figure of Goethe's Mephistopheles. Man has felt the "evil" as tempting. Why, we must ask, did he not think of God as "tempting"? If the devil represents distorted nature and God is primal, true nature, why does man feel so much more drawn to the devil than to God? Why the great, eternally frustrating effort to redeem man from "sin" (i.e., from the temptation of the devil), if the beauty, harmony, life-enhancing power of God is so obvious and so convincingly postulated?

The answer is, again: *the devil is tempting and so easy to follow because it represents the secondary drives that are so accessible. God is so boring and distant because it represents the core of life that has been made inaccessible by the armoring.* Therefore, God is the great unreachable goal and the devil is the omnipresent, engulfing reality. In order to make God a living reality, the armoring must be destroyed and the identity of God and primal life, of devil and distorted life, firmly and practically established. Unfortunately, God and the life process, which is nowhere so clearly expressed as in the orgastic discharge, are identical. Once this approach to God was blocked, only the devil could reign. And how he reigned! How tragic, how colossal this error of man, this endless search for the inaccessible experience of God, with the fatefully irrevocable landing in the devil's realm!

"God" as the representation of the natural life forces, of the bio-energy in man, and "devil" as the representation of the perversion and distortion of these life forces, appear as the ultimate results of character-analytic study of man's nature. With this conclusion, the basic task I have set myself in this book seems fulfilled. From here onward, orgone physics must take over. Now it is the problem of the ether that requires our strictest attention; it is the most basic problem of all physical theory and natural philosophy. However, the *character structure* of man, the observer of nature, and its biophysical core, the *orgasm function,* remain the guiding posts into this realm of nonliving nature. That must never be forgotten.

CHAPTER VI

COSMIC ORGONE ENERGY AND "ETHER"

It is not our objective here to prove the existence of an all-pervading ether; neither is it intended to prove the identity of the cosmic orgone energy and the postulated ether. All that is to be established at this point is the fact that there exists an all-pervading, observable and demonstrable energy. It is filling gaps in the comprehension of the universe, gaps that many generations of physicists and philosophers tried hard, but in vain, to fill with the concept of an all-pervading "ether" as the primal substratum of the basic functions in nature.

The time in which cosmic orgonomic functions have been studied is very short. It comprises not more than a decade. However, all observations within this short period have led to the following conclusion:

THERE IS NO SUCH THING AS "EMPTY SPACE." THERE EXISTS NO "VACUUM." SPACE REVEALS DEFINITE PHYSICAL QUALITIES. THESE QUALITIES CAN BE OBSERVED AND DEMONSTRATED; SOME CAN BE REPRODUCED EXPERIMENTALLY AND CONTROLLED. IT IS A WELL-DEFINED ENERGY THAT IS RESPONSIBLE FOR THE PHYSICAL QUALITIES OF SPACE. THIS ENERGY HAS BEEN TERMED "COSMIC ORGONE ENERGY."

First, let us summarize the general conclusions that follow from the fact that there is no empty space; and, second, let us summarize the phenomena that have forced upon us the conclusion that the primordial cosmic energy, hitherto postulated as the *"ether,"* has been finally discovered in a practical and concrete manner, accessible to direct observation and experimentation.

1. All physical theories resting on the assumption of "empty space" tumble if, and only if, the abstract mathematical structures that were to replace the concrete physical qualities of space cannot be reconciled with the new factual observations.

2. The qualities that characterize "space" must be of a strictly physical nature, observable and reproducible in high vacuum.

3. The theoretical supposition of an "ether" continues to be valid. The phenomena in the "vacuum" must agree with the qualities that had to be ascribed to the ether in order to explain the functions of field actions in space, such as gravity, light, attraction at a distance, "transmission of heat from the sun to the earth," etc.

4. The negative result of the Michelson-Morley experiment, which was designed to demonstrate the ether, must be comprehended.

The premises that led to the performance of the Michelson-Morley experiment rest on incorrect assumptions. Orgone physics starts from entirely new observations and new theoretical assumptions. Seen from a basic orgonomic viewpoint, reasoning itself must be understood as a function of nature in general. Accordingly, the results of mere reasoning must be secondary to observable functions of nature. As functionalists, we are mainly interested in the observable functions of nature; from there, we arrive at the functions of human reasoning by way of the emotional (bio-energetic) functions within observing man. As long as observable nature does not constitute the starting point for human reasoning, and, furthermore, as long as the function of reasoning itself is not deduced in a logical and consistent manner from observable functions of nature within the observer, all results of mere reasoning unsupported by observations are open to basic methodological and factual questions. This is clearly shown in the conclusion drawn by mere reasoning from the negative outcome of the Michelson experiment. Though I must leave a thorough critical evaluation of this experiment to the physicists, who are at home in the realm of its premises, the following remarks may be justified on the basis of some observations in orgone physics:

a) One of the premises of the Michelson experiment was the assumption that the ether is at rest; the earth, accordingly, moves through a *stationary ether.* This assumption is clearly proven incorrect by observation of the atmospheric orgone. If the "ether" represents a concept pertaining to the cosmic orgone energy, it is *not stationary, but moves more rapidly than the globe of the earth.* The relation of the earth's sphere to the surrounding cosmic orgone ocean is not that of a rubber ball rolling on stagnant water, *but of a rubber ball rolling on progressing water waves.* Thus the first assumption of the Michelson experiment becomes invalid.

b) Orgonomic observations make it essential to separate, within the function of "light," the "lumination" from the "excitation," which is propagated through space with the "speed of light." Light, accordingly, does not move at all but is a *local effect* of *orgone lumination.* Thus the second premise of the Michelson experiment becomes invalid, if one accepts—as one is forced to do—the clear-cut orgonomic observations in nature. I refer here to the orgonomic lumination effect in high vacuum, to the phenomena of "dawn," to the aurora borealis, to the corona of the sun, the luminating ring of Saturn, etc. If "light" is due to local orgone lumination and does not "travel through space" at all, it is quite understandable that in the Michelson experiment no phase

difference could be observed in the light beams which were
"sent" in the direction of the ether "drag" and perpendicu-
lar to it.

5. The physical functions in the vacuum must not contradict any cosmic functions at the basis of the planetary motions. On the contrary, they must, in due time, lead to an
integration of the function of the primordial cosmic energy
with the motions of the heavenly bodies.

6. There must be definite reasons why generations of
physicists and astronomers have failed to demonstrate the
ether in the strict sense of *physical* functions. These reasons
are to be found in the functioning of the *observer* and in the
method of human thinking itself.

We shall now proceed to summarize the more important
functions that have been observed and demonstrated in what
I termed "cosmic orgone energy" since its discovery in
bionous matter in 1936 and in the atmosphere in 1940.

1.   FORM OF EXISTENCE

Certain orgone energy functions can be demonstrated
wherever man is capable of directly observing nature and of
setting up the appropriate instruments that react to these energy functions, e.g., thermometer, electroscope, Geiger
counter, magnifying glass, darkroom lined with sheet iron,
and living organisms, be they protozoa or cancerous mice,
anemic human beings, or proteus bacilli.

ORGONE ENERGY CAN BE DEMONSTRATED EVERYWHERE
SINCE IT IS PRESENT EVERYWHERE. ACCORDINGLY, IT
PENETRATES EVERYTHING, THOUGH AT VARYING RATES OF
SPEED.

No arrangement has yet been found by means of which
one functioning realm of mass-free orgone energy can be

distinctly delineated from another, as one electric line can be sharply delineated from another. Therefore:

We must comprehend the living organism as an organized part of the cosmic orgone ocean, which possesses special qualities called "living"; we fail to understand this organism *bioenergetically* if we adhere to the mechanical energy potential. This mechanical potential, be it thermical, electrical or mechanical motion, is always directed from the higher to the lower, or from the stronger to the weaker system, and never vice versa. On the other hand, the living organism would not only *not* be able to keep up a much higher energy level as compared with the environment; it would also lose its heat, its motility, its energy into the surrounding, energetically lower, environment in a very short time. And the question would remain unanswered how it came about that such an organism could come into being in the first place. We cannot get around the fact that there is in nature another energy function, our so-called REVERSED, ORGONOMIC POTENTIAL; *orgone energy flows from the weaker or lower to the stronger or higher system.* This not only agrees with the basic functions of living organisms, but also can be directly observed in non-living nature, e.g., the function of gravity or the "growth" of clouds in the sky.

The orgonomic potential does not contradict the old mechanical potential. In fact, it explains how it is possible that a higher energy level can exist at all. It is true that, in accepting this function, the "second law of thermodynamics," the absolute formulation of the "law of entropy," becomes invalid. We know that many physicists feel uncomfortable with this law anyhow. And we have had to abandon many other such beliefs of absolute nature, e.g., the conservation of matter or the unchangeability of chemical elements.

The orgonomic concept of energy functions in the living

organism as it emerged from observation and deduction is
the following:

1. The living organism, as the stronger energy system,
draws its energy from the lower energy level: ORGONOMIC
POTENTIAL. This is valid not only for the organism as a
whole, but for the nucleus in each living cell, which draws
energy from the surrounding, energetically lower protoplasm.

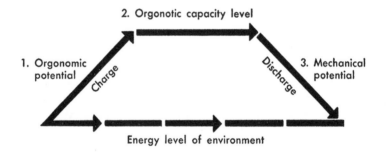

2. Orgonotic capacity level

1. Orgonomic
potential

Charge

Discharge

3. Mechanical
potential

Energy level of environment

Diagram of the ORGONE ENERGY METABOLISM *in living
bodies*

2. Each type or species of organism possesses its specific
energy level; it has a special "orgonotic capacity." Otherwise, the living organism would not stop accumulating energy and would burst or grow indefinitely.

3. All surplus of energy is discharged according to the
mechanical potential (from the higher to the lower level) in
mechanical movement, in orgastic convulsions, in radiation
of heat, etc.

4. There exists, accordingly, an orgone energy metabolism, a continuous exchange of energy in the cohesive
unit called organism. To summarize its main functions:
maintenance of a certain capacity level by means of *charging*
from the surrounding orgone ocean and from foodstuffs, and

by energy *discharge* into the surrounding energy ocean. The lower the capacity level, the weaker the capacity for charge, as in the shrinking biopathy. In the dying organism, the capacity to charge and to maintain the level of functioning is slowly lost. The capacity level sinks until it reaches the level of the surrounding orgone ocean. In the putrefaction process after death, the opposite of what went on during original growth occurs. The material tissues lose their cohesion due to loss of orgone energy; they fall apart; finally the unit disintegrates into bions and then into rot bacteria (proteus bacilli, etc.).

The main characteristic of orgone energy appears to be motion and, with it, metabolism. There exists such a thing as stoppage of orgone energy motion as, for instance, in severe cases of anorgonia. Such stoppage leads inevitably to a lowering of the capacity level and thus to final disintegration of the orgone unit called organism, as in death. I have been told that disintegration due to lack of orgone metabolism also occurs in wooden buildings that remain uninhabited over long periods of time. If we could find the reason why the capacity level of orgone systems is lowered after a certain period of functioning ("aging"), we would be able to approach in a practical manner the problem of how to lengthen life.

· ORGONE ENERGY IS PRESENT "EVERYWHERE," AND IT FORMS AN UNINTERRUPTED CONTINUUM. This continuum varies in different places with regard to its "denseness" or "concentration." We are still using these mechanical terms borrowed from the language of the physics of matter although orgone energy is not of a material nature. Therefore, we must be prepared to replace these terms with others that are more fitting to describe the functions of orgone energy. Orgone penetrates all space, including space occupied by solid matter. It penetrates a wall of cement just as it does a wall of steel. The difference lies in the speed of penetration: cement absorbs and discharges orgone energy slowly; steel attracts orgone energy strongly and quickly, but it also reflects it instantly, since metal seems incapable of holding orgone energy. This fact may have some bearing on the function of rapid energy flow through metal wires.

2.   MOVEMENT

The physical functions abstracted in orgone physics as "orgone energy" are always and everywhere *in motion,* or, expressed differently, *moving.* To this date, it has been impossible to ascertain an orgonotic condition that, with reference to another particular system, could be defined as "immobile" or "unchanging." A rock, which, in the physical sense, represents a certain material variation of the cosmic orgone energy, can be described as "resting" with reference to a second rock next to it; the orgone energy, however, which can be physically demonstrated in the rock, is never resting with regard to the same frame of reference.

Can the basic "law of the conservation of energy" be reconciled with the existence of an orgonomic potential? It probably can. A first tentative suggestion to this effect is offered in the following assumption: while some orgone units are forming in the orgone ocean by concentration, others terminate their single existence by energy dissipation into the orgone ocean. Thus, the energy lost by discharge or "deaths" of a number of orgone units would be picked up again to be concentrated in other units. The "running down of the universe" toward random functions would in this way be counteracted by new births of high energy potentials due to reversed concentration ("Creation"). The orgonomic (reversed) potential would make entropy unnecessary.

Orgone energy is basically of a dynamic, metabolic nature. This would seem to be true for any kind of energy, since energy is a function of motion, and vice versa. But classical physics speaks of "potential energy" as, for instance, that contained in water in a high basin. Nothing of the kind can be found in orgone energy; it never shows any condition that could be called static or immobile, except in its form of solid matter. It is this dynamic character of orgone energy that underlies the *functionalism* of all known orgone phenomena; this is true even for mechanical manifestations such as the sinus wave or free fall. Therefore, movement, dynamics, functionalism, changeability constitute specific, i.e., inseparable qualities of the cosmic orgone energy.

Within the framework of this motility, we can, by means of observation and experiment, discern many different kinds of motion:

a) *Wavy motions*

We can clearly see wavy, rhythmic motions over smooth water surfaces of mountain lakes and in the sky. This motility, too, is not uniform but varies continuously. There are no two parts of the surface of a lake that would have an equal motion at one and the same time. In addition, the wavy motion shows different rhythms at different times; the oscillations constantly embrace different areas. We search in vain for a mechanically uniform, static motion or condition. There is nothing within the realm of primal orgonotic functions that would appear as mechanical repetition. There seems to exist no law at all except the ONE of

b) *Pulsation*

ALL WAVY MOTIONS OF THE PRIMORDIAL ORGONE ENERGY PULSATE. It is necessary to distinguish the wavy form of orgone motion from its pulsation. The pulsation differs from the wavy motion in that

1) pulsation consists of alternatingly expanding and contracting movements, while the wave is a steady progression of wave crests and wave troughs;

2) in pulsation, the medium, e.g., the water in the lake,

seems to move from a certain center in all directions to and fro, while in waves the water swings up and down, thus marking the crests and the troughs of the progressing waves that run over the water's surface;

3) the pulsatory motion on the lake's surface moves slowly from west to east or not at all, while the waves run much faster in the same direction;

4) pulsation is an essentially *discontinuous* process, while wave motion is a *continuous* process;

5) pulsation is a function in three dimensions of space as manifested in the sphere-shaped propagation of radio signals. Waves, on the other hand, are, if examined singly, two-dimensional functions defined by wave length and frequency only.

If we follow the path of a certain wave crest or wave trough, we obtain a continuous line; the wave form itself is a continuous line. However, following the positions of extreme expansion or contraction in the pulsatory function, we obtain *points* and not a line. Through observation of the motion of atmospheric orgone at mountaintops, we can clearly discern *pulses* and *waves*. The pulsatory peaks are superimposed on the course of the waves in the following manner:

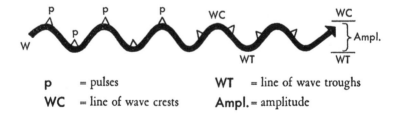

p   = pulses   WT   = line of wave troughs
WC = line of wave crests   Ampl. = amplitude

Schematic presentation of the differences between pulses (p) *and waves* (w)

My attention was first called to this basic difference between p and w in 1935 when I measured and photographed
bio-energy at the skin surface. It was not until 1948 that I
understood the inner functional interrelation between pulses
and waves in the orgonotic system. This was made possible
by utilizing orgonotic pulses in setting a spinner type of
motor into rotary motion.* In the original observation, the
pulses p were superimposed on the wavy motion of the organismic orgone energy like mountain peaks on a mountain
range:

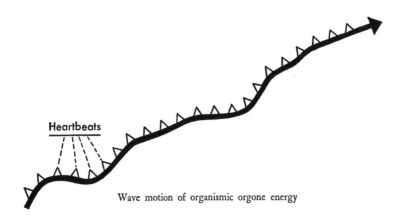

Heartbeats

Wave motion of organismic orgone energy

Whereas in the mechanism of a radio transmitter the pulsatory signals are, according to the theory, transformed into
waves synchronically, the pulses of the heartbeat are *not*
synchronized with the rhythm of the orgone waves. The
pulses are distributed regularly while the waves are in a state
of constant change. This is true for the living organism as
depicted in the sketch above. It also seems to be true for
the motion of the atmospheric orgone.

---

* *Cf.* Communication in *Orgone Energy Bulletin* 1, 1949, pp. 10–11.

From now on, we shall designate the symbol *p* for *pulses* and the symbol *w* for *waves*. These distinctly separate functions of the *one* basic function of PULSATION (*P*) will later be studied thoroughly in certain orgonometric functions. They express a most important relationship between *discontinuous* (p) and *continuous* (w) functions of nature. We shall have to find out how and to what extent these functions are related to Planck's quanta (*discontinuum*) and to classical wave mechanics (*continuum*).

As a preliminary measure, we may try to coordinate *p, w* and *P* functionally in the following manner:

PULSATION (P) ——————— pulses or impulses (p)
                        Waves (W)

Thus, pulsation would constitute the common functioning principle of both pulses and waves, the two varied functions of pulsation. It will require elaborate functional deductions to derive the energy formula of orgonotic primal pulsation from relevant cosmic functions. They will be given in a different context.

c) *The west–east movement of the atmospheric orgone envelope*

The west–east direction in the motion of the atmospheric orgone energy has great significance independent of the special variations that appear in the wave motion and in the pulses. The west–east direction is in agreement with the direction of rotation of the earth's sphere; thus, it is also in agreement with the general direction of rotation in the planetary system. A reversal of this generally valid direction on the earth's surface takes place only before thunderstorms or heavy rain occurring *to the west* of the observer. The following diagram illustrates the reversal of direction in the movement of the orgone energy envelope.

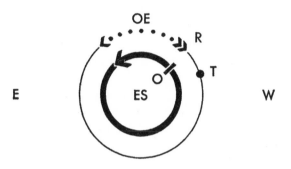

E = East	**W** = West
OE = Orgone envelope	**R** = Reversal to E → W
O = Observer	**T** = Thunderclouds
ES = Earth's sphere	‹‹•••••›› "Thinning" of orgone envelope

This reversal can be explained without contradiction by the attraction exerted upon the freely moving orgone energy at R by the highly concentrated orgone in the clouds to the WEST (to the west of the thundercloud no reversal will take place, since the attraction will act in the same direction as the general motion of the orgone envelope). It is hard to state definitely whether or not the strong west–easterly wind, which often arises after the clearing of the sky, i.e., after the return of the west–east direction of orgone motion, is a consequence of the filling of the gap (⟨ . . . ⟩⟩) that arose because of the reversal at R.

3.   VISIBILITY IN THE DARK
(AUTOGENOUS LUMINATION)

A room of at least 6 ft. sq., made completely light-tight and lined with sheet metal on the inside, is required to observe the orgone energy phenomena in the dark. After about 15 to 30 minutes' adaptation, the room appears BLUISH-GRAY and not black. Slow-moving, foglike formations are clearly discernible; the longer we sit in the dark metal room, the more distinct the light phenomena become. After some time, bluish-violet, strongly luminating light points appear. Later, when our organism has excited the orgone energy in the room to a sufficient extent, a "concentration" sets in the foggy formations; rapid, yellowish-white, lightninglike streaks of light ("Strichstrahlen") cross the room in all directions. We can magnify these whitish rays by observing them against an opaque, plastic screen with a magnifying glass (4x–6x). For a more detailed description of darkroom phenomena, I refer the reader to chapter IV of my book *The Discovery of the Orgone,* Vol. II., *The Cancer Biopathy.*

4.   CHANGE OF FORM

In 1939, when I first discovered and observed the orgone lumination in the darkroom, I believed that orgone energy consisted of three distinct forms: bluish-gray, foglike formations, deeply blue-violet luminating dots, and whitish, rapid rays. Since then, these three forms have been found not to be distinct types of orgone energy, but three forms of one and the same energy under different conditions. Orgone energy changes from the foglike to the raylike condition when it is excited or irritated. Such excitation can be stimulated by

a) *metallic substances;* metal does not hold or absorb orgone energy, but reflects it rapidly, thus constituting an "obstacle," if such a mechanistic term is applicable at all in the realm of orgone functions, i.e., ether functions.

b) *living organisms in the metal-enclosed darkroom;* the organismic orgone energy apparently excites the atmospheric orgone energy and vice versa.

c) *electromagnetic, discontinuous field action;* an induction coil system will speed up the change from the foglike to the raylike formation quite considerably in 20 minutes instead of the customary 1 to 2 hours.

Every type of condition and form is mobile, dynamic, differing in speed, and never static-mechanical.

5. LUMINATION

The visibility of orgone energy in the darkroom is obviously due to the function of lumination. Orgone energy "emits" or "develops light," or, expressed differently, it functions in such a manner that our visual sense perceives light; it luminates. This orgonotic lumination, no matter under what conditions it may occur, is in most cases of a *bluish*-gray, *bluish*-green or *bluish*-violet color. This distinguishes orgone lumination from other types of lumination such as that of gases; neon gas luminates red, argon white, helium green. Orgonotic lumination in a *vacuum* is distinctly bluish-violet and it produces a blue reaction on color film.

The lumination is strongest when there is an exciting contact between two orgone energy fields or between an orgone energy field and an electromagnetic field.

Orgonotic lumination is "cold"; no heat develops as, for instance, in the passage of electricity through thin wire or as in chemical combustion. Atmospheric conditions such as sheet lightning over wide areas of the sky, the wavy aurora borealis, and the bluish St. Elmo's fire, together with the soft glow of glowworms and the bluish lumination of wood that has disintegrated bionously, are examples of "cold" lumination functions of orgone energy.

Microscopically, orgonotic lumination in living cells and in bionous matter can be observed in the strong refraction of light. Strongly charged earth bions or red blood cells show a strongly luminating aura around the membrane. When these cells weaken and lose their orgone charge, this aura, or energy field, disappears.

6. HEAT PRODUCTION

Whereas orgonotic lumination is "cold," other functions of orgone energy are accompanied by a more or less marked rise in temperature. The heat level of living organisms is usually higher than the temperature of the surrounding air, and the temperature of the earth's atmosphere is constantly higher than the much lower temperature of surrounding space. At the orgone accumulator, a constant temperature difference (To—T) is maintained between the air directly above the accumulator and the air surrounding it. These differences vary, on the average, from 0.3° to 1.5°C in closed rooms; in the open air, they often reach high values such as 15° to 20°C in the sun. This temperature difference is apparently due to heat developed by the reflection or stopping of the kinetic energy of the orgone energy motion at the metal walls. Removal of the inner metal walls reduces the difference to 0° or nearly 0°.

According to the current orgone-physical assumption, the continuous difference To—T is a manifestation of a concentration of orgone energy in the orgonotic system, be it a living organism, the planet, or an orgone accumulator. It demonstrates the "orgonomic potential" from the lower to the higher level, and it contradicts the general, unrestricted validity of the second law of thermodynamics. There exists not only a process of dissipation of energy in the form of heat, but also the reverse process of the building up of energy.

7. "STATIC ELECTRICITY"

"Static electricity" is a specific function of the atmospheric orgone energy. Paradoxically, the theory of electricity assumes that the atmosphere is both *free* of electrical charges and *full* of static charges, "static" being of an *electrical* nature. Sheet lightning and thunderstorm lightning have remained unexplained. Orgonomy demonstrates the charges in the atmosphere by means of the spontaneous electroscopic discharge, the so-called natural leak in classical physics. This discharge is more rapid in a less concentrated orgonotic atmosphere and slower in a more highly concentrated one. It is, furthermore, slower at noon than in the early morning and evening; this fact contradicts the ionization theory (*cf. The Discovery of the Orgone,* Vol. II, *The Cancer Biopathy.*)

8. "CONCENTRATION" OF ORGONE ENERGY

Two-to-threefold concentration of orgone energy in the atmosphere is indispensable for many orgonomic procedures and experiments. Certain experiments do not succeed in the natural atmospheric concentration, such as, for instance, the charging in high vacuum. The necessary concentration is achieved by the presence of an "orgone room" or several orgone accumulators in the building where the experiments are carried out. The orgone accumulator is capable of concentrating atmospheric orgone energy by the arrangement of its layering. It consists of two or more (up to twenty) layers, each constructed of nonmetallic substance on the outside and sheet iron or steel wool on the inside. This arrangement influences the atmospheric orgone energy in such a manner that its movement toward the closed space is greater than toward the outside. An "orgonomic potential" is created from the lower level outside toward the higher level inside and is continuously maintained; the orgonomic potential can be demonstrated by the slower discharge of electroscopes on the *inside* and by the constant temperature difference *above* the upper metal plate (To—T).

Concentrated orgone energy has many beneficial effects on living organisms, which I have tried to describe in my book *The Cancer Biopathy.*

9. THE ORGONOTIC POTENCY AND SENSITIVITY OF THE OBSERVER

Human organisms with low orgonotic potency or severe armoring do not perceive orgone energy phenomena easily, in contradistinction to healthy organisms. People suffering from a cancerous shrinking biopathy do not feel the orgone in the accumulator until after several days or even weeks, i.e., until they have become sufficiently charged. Orgonotic perception, too, is weakened in observers with low orgonotic potency. Such experimenters will, for instance, not be able to produce lumination in a gas tube. Furthermore, they will have difficulties in discerning the light phenomena in the dark and they will be unsure about the sensations of heat and prickling that the well-charged organism perceives so easily.

The biophysical structure of the observer is, therefore, of the utmost importance in orgonomic work. Severely armored individuals easily develop anxiety in the darkroom when they become aware of the lumination. On occasion they react with panic. They will try to interpret away the phenomena with empty words such as "It's only subjective," or "It's mere suggestion," or similar statements.

The structure of the observer is of importance since it is the organismic orgone energy in his sense organs that reacts to the external orgone phenomena. The inclusion of the structure of the observer in the judgment of natural phenomena is a very important, if not decisive, step forward toward the integration of the *subjective* and the *objective*, the *psyche* and the *physical*. It is chiefly the ignorance of the biophysical and depth-psychological functioning of the observer that has led mechanistically oriented scientists into the dead end street where theoretical physics finds itself today. These scientists, who otherwise have demonstrated such an excellent critical sense of inquiry, are still adhering to old, outmoded psychological thinking. They cling to a "consciousness" that somehow has no basis in the organism, no rooting in biophysical processes. They are unaware of the great progress that has been made during the first part of this century in connecting the functions of perception with the functions of the emotions, and in connecting the emotions with *bio-energetic*, i.e., truly *physical*, processes in the observing and reasoning organism. Natural-scientific research is an activity that rests on the interaction between observer and nature, or, expressed differently, between orgonotic functions *inside* and the same functions *outside* the observer. Thus, *the character structure and the senses of perception in the observer are major, if not decisive, tools of natural research.* We do not doubt for a moment that the physiological structure of a surgeon plays a decisive role in the operation he performs; or that the sense of balance and rapid movement in the flier are decisive for the mastery of aviation. But, in natural research, this principle has been badly neglected and misinterpreted. I fear that it is the emotional and, within the emotional realm, especially the *biosexual,* functions that have kept the classical researcher from closing the gap between observing (biopsychic) and observed (biophysical) nature.

In the realm of non-living nature, too, orgonomic research finds its best orientation when it adheres to what it has learned in the realm of living nature about the intensity of organ sensations, about the function of the orgastic convulsion, about the endoptic phenomena, about biophysical field reactions beyond the skin surface of the organism. The unimpeded orgonotic functioning in the observer and the experimenter is, therefore, a requirement of the very first importance in orgonomic natural research. It is regrettable that the thinker in classical physics does not find his way toward the understanding of the bio-energetic functioning in the observer and that he is inextricably bound up with the view of a phenomenological psychology of some fifty years ago, in which "consciousness" and "reason" are still floating freely, unrooted, unconnected, and uncomprehended in "empty space."

I wish to summarize the functions that have been attributed to the ether by many researchers who tried to describe the general substratum of all known physical phenomena. It is amazing to find that most of these functions, which have never been observed directly, coincide with many functions of the cosmic orgone energy, which have been observed directly and reproduced experimentally. The biopsychiatric problem that amazes us is this:

Observers of nature have described the cosmic primal energy correctly as far as its main functions are concerned. Still, they have been unable to make contact with these functions except by deduction; direct observation and experimentation with the ether were closed to them. It is obvious that this fact is not to be attributed to the ether but to the observer. Therefore, the problem is a biopsychiatric one. It is mainly concerned with the biophysics of perception, the interpretation of sense impressions and organ sensations. As the whole development of orgonomy has so clearly demonstrated, there was only ONE access to the physical study of the ether, the orgonotic current in man, or, expressed differently, the "flow of the ether" in man's membranous structure. For many ages, religion has called this primal force "God." We begin to understand now why most of the great physicists who dealt with cosmic problems, especially those of the ether, such as Newton, were occupied so intensely with the problem of God.

SUMMARY OF FUNCTIONS

As required for the ETHER	*As observed in* COSMIC ORGONE ENERGY
1. Exists universally.	Exists universally; orgone accumulators operate everywhere.
2. Fills all space.	Fills all space; orgone energy can be demonstrated in a vacuum.
3. Penetrates all matter.	Penetrates all matter.
4. Is the source of all energy.	Varies and manifests itself as heat, "static electricity," thunder- and sheet lightning, electricity, magnetism, gravitational attraction.
5. Changes into matter or mass.	Superimposition of two or more orgone energy waves ("Kreiselwelle") results in a mass particle.

ETHER	COSMIC ORGONE ENERGY
6. Is responsible for the cohesion of atoms.	Keeps unit of matter, bions, together. Cohesive force become free and demonstrable when solid matter disintegrates into bions = orgone energy vesicles.
7. Transmits light.	Transmits orgonotic *excitation* with the "speed of light"; "light" is a manifestation of orgonotic lumination and is of *local* character.
8. Is transparent.	Is transparent; can become visible as "refraction of light," as "heat waves" and "bad seeing."
9. No heat in ether.	Most orgonotic functions are "cold": lumination, movement through wire, attraction. However, reflection by metal creates heat, as does highly mobile concentration within matter, planet, organism.
10. No loss of energy.	No loss of energy; however, there is an "energy metabolism": a) flow toward higher level; b) maintenance of higher level, "capacity"; c) discharge toward lower level.

11. Is resting, stationary; earth moves through ether like a rotating ball on stagnant water.

Is always in wavy and pulsatory motion; the orgone envelope moves *more rapidly* in the galactic orgone ocean than the earth's globe; analogy is that of a ball rolling on water waves more slowly than the waves.

12. "Cannot be demonstrated"; misinterpretation of Michelson experiment was due to assumption that the ether was stationary and that "light" traveled through space.

Is clearly demonstrable everywhere visually, thermically, electroscopically, with Geiger counters; accounts for phenomena in nature hitherto unexplained: "natural leak," "bad seeing," "field action in empty space," "static," "cosmic rays," blueness of sky, ocean, distant mountains, "ionized cosmic dust" in aurora borealis, etc.

COSMIC SUPERIMPOSITION

O man! Take heed!
What saith deep midnight's voice indeed?
"I slept my sleep—,
"From deepest dream I've woke, and plead:—
"The world is deep,
"And deeper than the day could read.
"Deep is its woe—,
"Joy—deeper still than grief can be:
"Woe saith: Hence! Go!
"But joys all want eternity—,
"—Want deep, profound eternity!"

NIETZSCHE

CHAPTER I

STAGE AND MEADOW

The basic interest in the subject of this publication is human and not primarily astrophysical. *In what manner is man
rooted in nature?* is the question around which the theme
revolves. It is doubtless the orgone energy function in man's
reasoning that touches on reality.

The character structure of man, the frozen history of the
past four to six thousand years of human society, will determine man's fate and conditions in the near future. Looking
forward through a dense fog, which has obscured man's
view for several decades now, the author has tried to draw
the ultimate consequences from what he has learned about
human functioning over a period of more than thirty years
of intimate knowledge of the characterological backstage of
the public scene. Very little of the actual drama of present-
day social struggles will appear in these pages, however. The
author did not intend to study the impact of the backstage
events upon the performance on the public stage. On the
contrary, he has opened the door that leads from the backstage of the theater to the spacious fields and meadows
surrounding the theater of present-day human affairs. Observed from these meadows, under glittering stars in
endless heavens, the show on the stage appears strange.
Somehow, the endless heavens on silent nights do not seem
in any sort of accord with the show inside the theater or

with the subject of the performance. All that belongs to the show seems far off, unreal, and very much out of place if seen from *outside* the theater building.

Why does man present gay or tragic or pornographic love stories on the stage, with people crowding into the auditorium to laugh or cry or to shudder with lust, while deep in the woods that surround the meadows, policemen are busy disturbing lovers in silent, quivering embrace? It does not seem to make good sense.

This is only one small, insignificant example of the great discrepancy and the varied nonsensicalities in man's existence. We shall not delve into any of these social, psychological, biological or political issues, which have been thoroughly elaborated by the author in previous writings. The social problem does not seem to yield to any kind of inquiry within the sphere of man's thought and actions during the past few thousand years. Let us therefore try to look at it from outside.

The impetus to the present study came from some disturbing experiences in the Orgonomic Infant Research Center, which was founded by the author for the purpose of studying nature in the newborn infant. Orgonomic research had broken down completely the boundaries between the *bioenergetic* and *astrophysical* realms, heretofore kept strictly delineated by mechanistic natural science and transgressed only in mystical experiences, in a factually useless way. The newborn infant appears as an energy system that brings some definite cosmic laws of functioning into man's realm of operation, i.e., to remain with our analogy, with the infant, definite cosmic functions pass through the door that leads from the meadow and the open fields into the theater building and onto the stage of human drama.

In this respect, the newborn infant is comparable to the experience one often has when working with orgonotic pulses on the Geiger counter or on the oscillograph. One can easily switch over from pulses in the living organism to the same type of pulses in the atmosphere. One operates in a practical manner with the *common functioning principle,* the CFP, of man and cosmos. There is no longer any barrier between a human organism and its cosmic environment, which, of necessity, is and always has been its origin. One forgets the show on the stage and concentrates on this amazingly practicable identity of living and non-living functions.

On the human stage, it is forbidden by law under punishment of fine or imprisonment "or both" to show or even discuss the embrace between two children of the opposite sex at the age of three or five. Somewhere in the audience sits a human being, broken in his emotional security, full of perverse longings and hatred against what he has lost or never known, who is ready to run to the district attorney with the accusation that children are being misused sexually and that public morals are being undermined. Outside on the meadow, however, the genital embrace of two children is a source of beauty and wonder. What drives two organisms together with such force? No procreation is involved as yet, and no regard for the family. Somehow this drive to unite with another organism comes with the newborn when it passes from the meadow onto the stage. There it is immediately squelched and smolders under cover, developing smoke and fog.

Inside, on the stage, the embrace between two children or two adolescents or two grownups would appear dirty, something totally unbearable to look upon.

Outside, under the glimmering stars, no such reaction to the sight of the embrace of two organisms would ever occur in sane minds. We do not shudder at the sight of two toads or fish or animals of other kinds in embrace. We may be awed by it, shaken emotionally, but we do not have any dirty or moralistic sentiments. This is how nature works, and somehow the embrace fits the scene of silent nights and broad meadows with infinity above. The intellectual cynic and the smutty barroom hero, of course, belong on the stage and not in the meadow, where they would certainly disturb the harmony and not fit into the picture. But we would refuse to believe that a meditating Indian sage would object to the sight or not fit in.

Somehow, the deeply searching human mind has never failed to find itself in the meadows of nature, outside the theater of human stage shows, be it on high mountains or beside blue lakes. Somehow, the harmony in natural functioning belonged to the sage. It does not matter here whether or not human meditation has ever succeeded in lifting the veil. It has at least tried to do so and always outside the realm of human stage performances, be it a theater, a political gathering or a religious ceremony. When Christ found himself in trouble, he went to meditate completely alone on a meadow or a hill, in silent spaces. And again, something important, though inscrutable, was brought back from the meadow or the mountain onto the human stage.

Every single religious movement in the history of man has tried to bring the message of the emotional depth from the meadow onto the stage inside, in vain.

Tolerance, goodness, patience, brotherhood, love and peace are, as elements of this mood under glittering stars, contained in every religious creed; but the moment they were brought to the inside of the theater and onto the stage, they became a farce and a sham. Why?

Astronomy has always been in close touch with this same mood. Kepler brought the idea of a living force, the *vis animalis,* that governs the heavens as it governs the living organism right onto the stage. It did not survive.

The constellations of the stars in the heavens in ancient times were represented, most fancifully, by different living creatures—the scorpion and the bear, Andromeda and Hercules and Pisces, etc. Thus, man knew that somehow he came from the heavens into which, in nearly every religion, he believed he would return after death on earth.

For ages, man has projected his own image onto the heavens in the shape of different gods in human form, again showing that he believed himself to be somehow rooted in the heavens.

In the belief of the return of the soul, of reincarnation (and the believers have not been simple fools, as the dried-up creatures on the political stage want us to believe), man has somehow searched for a reality in which to root himself in the vastnesses of the universe. So far, in vain!

In recent times, more and more human thinking has come to assume that the idea of a universal natural law and the idea of "God" are pointing to one and the same reality.

Abstract mathematics, from the Pythagoreans to the modern relativists, has somehow assumed that the human power of reasoning is closely related to cosmic functions. True, no concrete links between reason and universe became evident. Still, the close connection was taken for granted. Mere reasoning seemed to have corroborated such a close interrelation between "mind" and "universe." However, it is not readily comprehensible what these links are. Orgonomy has contributed some major insights into this riddle by disclosing the transitions from reasoning to emotions, from emotions to instincts, from instincts to bio-energetic functions, and from bio-energetic functions to physical orgone energy functions.

Thus, the impelling force to search and the religious belief meet somewhere in the vast spaces. But both reasoning and belief instantly distort the clarity of the meadow experience when they transfer it onto the human stage. WHY? Is it because man is a different being on the meadow from what he is on the stage? Probably, but the answer is not good enough.

Now, the boundaries separating religious belief and pure

reasoning have been crossed, or rather wiped away by orgone research. It was shown in *Ether, God and Devil* that both reason and belief are rooted in the orgonotic, bioenergetic functioning of man. They are both rooted in one and the same functional realm.

Thus, it appears that all the events on the stage are somehow rooted in events on the meadow. But the common root is obfuscated by definite changes that occur during the passage through the door leading from the vastnesses of nature to the narrowness of the stage. Outside, everything seems to be ONE. Inside, the stage proper is cleanly separated from the auditorium. Outside, you can appear as you are. Inside, you have to disguise your true appearance by a false beard, or a false pose or a make-believe expression. Outside, two children in deep embrace would not astonish or shock anyone. Inside, it would immediately invoke police action. Outside, a child is a child, an infant is an infant, and a mother is a mother, no matter whether in the form of a deer, or a bear, or a human being. Inside, an infant is *not* an infant if its mother cannot show a marriage certificate. Outside, to know the stars is to know God, and to meditate about God is to meditate about the heavens. Inside, somehow, if you believe in God, you do not understand or you refuse to understand the stars. Outside, if you search in the heavens, you refuse, and rightly so, to believe in the sinfulness of the natural embrace. Outside, you feel your blood surging and you do not doubt that something is moving in you, a thing you call your emotion, with its location undoubtedly in the middle of your body and close to your heart. Inside, you do not live with your total organism, but only with your brain; and not only is it forbidden to study emotions, more, you are accused of being an adherent of phrenology and mysticism if you experience emotions in the same way inside as you do outside. Outside, there is such a thing as the movement and quivering of everything, from

the atmosphere to your nerves; inside, there is only empty
space and atoms dissolved into an endless row of "particles."

Let us stop now. It is enough to have shown the great
discrepancy.

We are now moving into the open spaces to find, if possible, what the newborn infant brings with it onto the human
stage. This study will, among other things, deal with hurricanes, the shape of the galaxies, and the "ring" of the aurora
borealis. This will astonish many a reader. What, he will inevitably ask, has a well-known, distinguished psychiatrist to
do with hurricanes, galaxies, and the aurora borealis? Is not
this proof enough of the rumor that he went "off the beam"
some years ago, after having reached a high degree of distinction in the field of psychiatry? It is not the writer who
went "off the beam," but the reader who thinks that way. He
has forgotten his origin and refuses to be disturbed in the
enjoyment of the stage show make-believe.

He has refused to leave the theater and to follow us
through the door onto the vast meadow whence all being
stems. He has not realized that a newborn infant cannot
possibly be understood from the viewpoint of a culture into
which it is being born. This is its future. It can only be understood from where it came, i.e., from OUTSIDE the stage.

Hurricanes, galaxies and the aurora borealis come into
the view of a human being who deals with the mentally sick
and with newborn infants if he follows consistently the red
thread of inquiry and reasoning that leads outward from un-
hampered observation of man's behavior toward his origin in
the cosmic realm of functioning. Those who wish to stay inside and refuse to move out are, of course, entitled to do so.
But they are not entitled to pass judgment on the experience
of others who do not believe in the rationality of the stage
show, who refuse to accept the dogma that what man displays inside the narrow space on the stage is his true being
and his true nature. Those who remain sitting in the tight

little place have no right whatsoever to judge what the wanderer on the outside experiences, sees, smells, lives through. No dweller on 32nd Street who never left New York would dare pass judgment on a report from an explorer of the North Pole. Yet, without ever having cared even to peep outside through the keyhole of the door, he usurps the right to pass judgment on the experiences of orgonomy, which operates far outside his narrow, tight, little stage. Let him be modest and confine himself to his own little world. We do not permit him to have opinions, and with a show of authority to boot, about things he never dreamed of approaching. He may be an authority on the stage of the theater, or a well-trained critic of the play, or he may be an actor playing the role of a professor of biology or astronomy. But in all these cases he is within the theater building. And unless he actually steps outside onto the meadow and looks around himself, seeing what is to be seen there in the open spaces, he had better be quiet and remain sitting comfortably where he is. Nobody will blame him. Outside, however, he is no authority whatsoever. There are no false beards outside, only living beings searching and wondering about where they came from and why they are there. We shall be glad to take his hand and lead him out into the night, where we have learned first to see and to feel what we intend to measure. We shall be happy to do so. But first let him remove his false beard of dignity. Let him be a man first.

Finally, it should be clearly stated that the seeming immodesty of the scope of this investigation is a quality of the function "cosmic superimposition" and not of the investigator. We are dealing with cosmic dimensions to be measured in "light years," not in seconds.

CHAPTER II

SURVEY ON MAN'S ROOTS IN NATURE*

The serious student of orgonomy is now invited to enter an airplane and fly high over the territory that has been made accessible through the discovery of the cosmic orgone energy. We are leaving behind us the mechanisms of distorted human nature, the biopathies and the neuroses, the miseries of infancy and the agonies of adolescence, the political irrationalism, as well as the production of goods. We shall survey the land where no human foot has ever trod, where no security but only functioning exists. The survey should serve one purpose only: to prepare for future possible settlements in a new, unknown territory. We shall survey our future home of astrophysical knowledge.

The existence of the new territory now to be surveyed became known not through the study of matter or mechanical movements but of man's basic emotions. To the mechanistically or chemically oriented mind this sounds rather queer. What, it asks, has the ecliptic, the yearly path of the sun, the aurora borealis, or a hurricane to do with human emotions? Mystical distortion of true knowledge is suspected. To this, in preparation for our flight, we answer: *It is always a sign of ignorance or of a mystical orientation to put man and his emotions outside the pale of physical nature.* Man is a part

*Elaboration of a lecture given at the Second International Orgonomic Convention at Orgonon, Rangeley, Maine, on August 26, 1950.

of nature; he grew out of natural functions. It cannot possibly be otherwise. It follows from mere reasoning about natural evolution. There is no valid counterargument to this statement. Man, including his emotions, evolved from nature as one of its developmental products. Once this conclusion is accepted, the next question follows: IN WHAT MANNER IS MAN ROOTED IN NATURE?

The chemo-physical base of operation, also, views man as being rooted in nature. There, it is the chemical elements and the electrons that connect man with nature. All medicine and all education of the last century were based on the chemophysical rooting of man in nature. However, the mechanistic-materialistic viewpoint was incapable of including human emotional life; thus, mystical and spiritual dogmata filled the gaps. Here, as is well known, the spirit, the soul, the "something" within man that feels and cries and laughs and loves and hates appeared to be connected with an immaterial world spirit; it represented in more or less clear terms man's connection with the creator of the universe, with "God." Thus, mechanics and spiritualism supplemented each other, with no bridge between the two realms. We had, accordingly, a science of physical nature and a science of moral conduct, or ethics.

Education, medicine, government, etc., were all geared to this dichotomy in man's existence. In education there were the good, God-fearing, and the bad, devil-inspired, children; in medicine one injected calcium, vitamins, sulfa drugs, or one applied the knife to the frontal lobe in cases of emotional disorders; in government this view has led to the establishment of God-sent, absolute monarchs or fuehrers who wield full mechanical as well as spiritual power over men. In natural science the dichotomy prevailed in the form of atoms here and complete ignorance of and disregard for the emotions there; the result of it all was an "empty space at rest" and cosmic equations that cleanly resolved to zero. The

Newtonian and the Goethean view of nature remained irreconcilable. The best among the physicists of the twentieth
century have given up hope regarding their own structure of
thought; they, too, are looking for the new land. Lecomte du
Nouy writes:

Physicists of the 19th century had drawn a picture of the universe
that was as satisfactory and reassuring as today's picture is unsatisfac-
tory. There is the same difference between our science and that of our
grandparents as there is between a cubist or surrealist painting and a
Meissonier or a Whistler. The small indivisible balls, which we fondly
dreamed represented atoms, gave way at first to minute solar systems
in which the electrons were the planets. To explain the discontinuity
of energy, it then became necessary to allow the electrons to jump from
one orbit to another. At that time they were considered as particles of
matter but with a mass dependent on their velocity, which was most
disturbing. When moving to an outside orbit a quantum of energy was
absorbed; when passing from the outside to inner orbits a quantum was
emitted. It was admitted that eight electrons could occupy an orbit.
The central nucleus—the "sun"—1840 times heavier than the electron,
carried a positive charge that maintained the electrons (negative electric particles) on their orbits. This model was certainly not ideal and
raised many difficulties of detail (for example, the rotation of an
electron on an orbit was supposed to entail neither absorption nor emis-
sion of energy, which is not very clear). But it had become familiar
and in spite of its complexity we considered it as a friend; we had begun
to forget its imperfections. It was, after all, "conceivable," and there
was something reassuring in the fact that there existed only one ultimate element that was the same for matter and electricity. We had no
sooner become accustomed to it, I might almost say attached to it, than
we learned rather brutally that this atom was only an impostor and
that the real atom had never resembled such a monster. *We were told
that there were not only two elements, the electron and the proton,
but at least three, one positive and one, the neutron, which carries no
charge; the mesons, positive and negative; the photon, quantum of
light, which like the particles is constituted of energy, and two entities
whose reality is limited to the necessity for balancing equations, the
neutrino and antineutrino, actually bookkeeping particles. Furthermore, only one electron can occupy an orbit, and today we can hardly
even speak of an orbit. We cannot even talk about an electron, in the*

sense that we did a few years ago, for the electron is at the same time a particle—perhaps deprived of mass—and a wave. Strictly speaking, it is not even a particle; it is only the expression of the probability that the properties that we attribute to the electron are to be found in a certain point of space. To be clearer, we can say that the electron is a wave of probability. The current notions of time and space no longer apply to these entities, which evolve in a pluridimensional space. (Italics are mine.—W.R.)

The Road to Reason, by Lecomte du Nouy

What has orgonomy to offer here? Does it collect the remaining fragments of a shattered old world picture that went to pieces or does it start anew?

It starts from scratch, in a basic, fundamentally independent fashion, without borrowing theories from classical science. Not because it wants to but because it has to. Its point of departure is not the electron nor the atom; it is not a linear motion in empty space, nor is it a world spirit or an eternal value. Its point of departure is the observable and measurable functions in the cosmic orgone ocean; from which all being, physical as well as emotional, emerges. Man, from this viewpoint, is, together with all other living beings, a bit of specially organized cosmic orgone energy.

Obviously, what constitutes man's roots in nature is not what distinguishes him from nature at large. Thus, man is not rooted in nature by his ability to talk, think, walk, eat, nor by the chemo-physical components of his physical structure such as salts, water, sugar, carbohydrates, etc. It is surely not his social organization or civilization that connects him with nature. These functions are variations apart from basic nature. Nature does not walk or think, talk or eat, nor is it composed of proteins, carbohydrates and fats. The common functioning principle that unites man with nature is something entirely different from all these things, something totally unknown.

Socio-economic philosophy has dealt with only one of the many deviations of man from nature, the economy of goods produced by tools used by man. The tool is a specifically human creation. Orgonomy, on the other hand, roots man in nature in the common functioning principle (CFP), i.e., *in functions man has in common with basic natural functions.* Since the CFP is always broader than later variations, the orgonomic viewpoint is much broader as well as infinitely deeper than the economic point of view.[1]

Nature, before life emerges from it as a special variation, has no economy, does not propagate, divide, walk, talk, eat, or perceive. Which functions, then, govern basically both non-living *and* living nature? Where is the red thread that runs through all of it, from the primordial orgone unit, visible in the darkroom, to the highest manifestations of life in man? A bold and terrifying question this is, indeed. However, we must not shrink from it. On the perfect formulation of this question much of future functional natural research depends.

When the going in unknown territory becomes rough, when the view blurs and confusion impends, it is necessary to return to realms of well-established knowledge. The red thread that hitherto guided our total research was found in the orgasm function. It was discovered that the orgastic convulsion governs all of the animal kingdom at the very roots of its bio-energetic existence. It was, furthermore, ascertained that the four-beat—*tension→charge→discharge →relaxation*—also governs cell division. Expansion and contraction, the two basic paired functions of the orgasm, also dominate the development of the embryo. In addition, this same function is clearly visible in the behavior of protozoa such as vorticellae, etc.

Does the orgasm function, as formulated here orgonometrically, also permeate non-living nature? Orgonomy answers this question in the negative. It assumes that the or-

---

[1] *Cf.* Reich, "Orgonometric Equations, I," *Orgone Energy Bulletin,* October, 1950, pp. 161–183.

ORGASM $\dashv <$

        EXPANSION $\dashv <$ TENSION / CHARGE

        CONTRACTION $\dashv <$ DISCHARGE / RELAXATION

gasm function in the formulation given above governs only the entire living realm and that non-living nature does not show the sequence: expansion → contraction.

Is not the earthquake or the thunderstorm an event similar to the orgasm in the animal world? We must not yield to such appealing analogies. It is true that in a thunderstorm and in an earthquake a tension builds up and is released by a discharge of energy. The analogy goes far indeed, and many a poetic mind has delved extensively into it. However, careful scrutiny of the analogy between the animal orgasm and cell division on the one hand, and a thunderstorm or an earthquake on the other hand refutes a functional identity. The question is: does a thundercloud constitute an "orgonotic system"? Obviously, it does not; it possesses no "core," no "peripheral membrane," and no "energy field." It is not "organized" like a living system. Therefore it does not convulse, it only discharges accumulated charges.

It is more difficult to refute a functional identity between the orgastic discharge in a living organism (including cell division) and an earthquake. We are dealing with an "orgonotic system" in both cases; for the earth's globe also possesses a core of energy, a membrane (the earth's crust), and an orgone energy field, the "orgone envelope." But does the planet convulse like a living organism? We must not mistake a dislocation of parts of a system for convulsion. The convulsion in a living organism is a total event that not only does not threaten the integrity of the system, it enhances its well-being and constitutes, as an integral physiological part of the whole, a basic function of the energy metabolism. No such function is discernible in the earthquake. It is more akin to the explosion of an overheated boiler than to an orgastic discharge. The analogy does not work. Thus, we must conclude that the orgastic convulsion is specific for the living domain only; that it differentiates the living from the non-living.

Where, then, is the functional identity between the non-living and the living to be sought?

In the chapter on the "Expressive Language of the Living" (*Character Analysis*, 3rd edn.), it has been suggested that man's orgastic longing is somehow pointing toward cosmic functions. No answer was given or attempted there. However, it was pointed out and emphasized that the orgastic longing of man, including all its disguised expressions such as mystical ecstasy, cosmic longing in puberty, etc., seems directed toward a basic function that precedes and induces the orgastic discharge: SUPERIMPOSITION.

Fig. 1. Superimposition and fusion of two living orgonotic systems

The longing for the genital embrace is profoundly expressed in the belief in a "universal spirit," in "God," the "creator." In basic natural science it is revealed in the search for the "natural law."

The function of sexual fusion is taken too much for granted to arouse curiosity about its place in the general course of natural events. Yet, to the searching mind it poses a stunning riddle: *whence stems the overpowering drive toward superimposition of male and female orgonotic systems?*

This question, far from being futile, turns out to be the key to a number of major riddles in astrophysics.

However, since man has banned the subject of the superimposition of two organisms from his scientific thinking in all his universities of higher learning, he has missed the approach to a great number of basic astrophysical functions and has become ensnared in an insoluble, rigid antithesis between sex and morals, nature and culture, bad and good, devil and God.

In *Character Analysis* and in *Ether, God and Devil*, it was shown how man runs away from his deepest core of bio-energetic existence and how strongly he protects himself against perception of this core. Man's biophysical armoring provides the explanation for the fact of the great runaway as well as the reason for the evasion of the basic questions of his whole life—his religion, his natural philosophy and, last but not least, his quest for knowledge about nature. Man must not perceive or understand his own living core; he must keep it secluded and inaccessible if he wishes to maintain his present social organization. *The great misery in which he finds himself entangled is due to his armoring, which cuts him off from his great bio-energetic possibilities and potentialities.*

After thus having established the HOW of the great runaway, we encounter the next question: *Why* DID HE START RUNNING AWAY IN THE FIRST PLACE?

CHAPTER III

THE FUNCTION OF SUPERIMPOSITION

The sexual embrace, if abstracted and reduced to its basic form, represents *superimposition* and the *bio-energetic fusion* of two orgonotic systems. Its basic form is the following:

Fig. 2

We have learned to reduce form to movement. Form, in orgonomic functional thinking, is *frozen* movement. Ample evidence has indicated that superimposition is due to bio-energetic forces functioning beyond voluntary control. The two orgonotic systems involved are driven to superimpose by a force that, under natural conditions, i.e., not restricted by outer or inner hindrances, is beyond their control. It is *involuntary* bio-energetic action. Basically, this function cannot be stopped, just as the heartbeat or intestinal peristalsis cannot be stopped, except by forceful interference or by death. When two children of different sexes, three to five years old, superimpose[1] and their organisms fuse orgonotically, we are not dealing with propagation, since no new individual will result from this fusion. Neither are we dealing with the "quest for pleasure" in the psychological sense. The pleasure involved in superimposition is the experiential result, and not the driving force of the act. Let us forget for a moment all the complicated higher functions that later are added to natural superimposition. Let us reduce it all to functioning beyond the individual and even the realm of the species. Let us penetrate deeply enough to see this function as an energy process that runs a certain course quite autonomically and with unequivocal effect. If we do this, then we clearly see in it a trans-individual event, something that takes charge of life and governs it.

Further careful observation tells us that bio-energetic superimposition is closely linked with plasmatic excitation and sensations of current in two orgonotic systems, be they children, adolescents, or grownups. It is absolutely necessary, in order to visualize this function in its proper aspects, to abandon all the many social, cultural, economic, psychological, and other implications that, in the case of man, have complicated and all but obliterated its original, bio-energetic functioning.

Reduced and abstracted in its purest form, superimposition in the biological realm appears as the approach through attraction and full bio-energetic contact of two orgonotic streams. Membranes, organs, fluids, nerves, will power, unconscious dynamics, etc., must be discounted here, since they do not constitute superimposition. Superimposition of two orgone streams appears as a common functioning principle (CFP) of nature that fuses two living organisms in a specific manner—specific to the basic natural function, and not to

---

[1] Cf. Reich: "Children of the Future, I," Report on the Orgonomic Infant Research Center, *Orgone Energy Bulletin* (October, 1950), pp. 194–206.

the two organisms. In other words, *superimposition of two orgone energy streams reaches, as a function, far beyond biology.* It governs other realms of nature, too, as it governs living systems. In order to find out which realms of nature beyond the living realm are governed by superimposition of two orgone energy streams, we must not deviate from its basic form and movement. Orgonometrically abstracted, it is this:

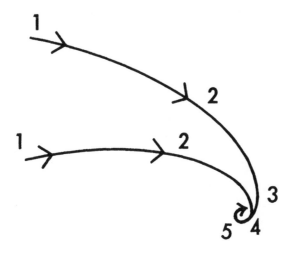

Fig. 3. Basic form of the function "superimposition"

Its functional characteristics are:
1. Two directions of energy flow.
2. Convergence ("attraction") and mutual approach of the two energy streams.
3. Superimposition and contact.
4. Merger.
5. Sharp curving of path of flow.
Finding superimposition in realms of nonliving nature would be a first decisive step toward finding a cardinal root

of man in nature, a *common functioning principle* that, already present and working in nature at large, also permeates in a basic fashion the animal kingdom, including man.

The following is a sweeping generalization. It was pointed out at the very beginning that what we are doing here is no more than flying high above a vast territory, the exploration of which will require painstaking, detailed efforts. We are free later on to abandon parts of it or the whole aspect, should it resist strictest observational and experimental as well as orgonometric scrutiny. We are also free to construct the framework of a future detailed operation, to retain its general features, its layout, and its basic characteristics while changing most of its inner detailed constitution. We are free to leave the confirmation or refutation of this construction to others. However, we would have to remind anyone who would approach a task of such magnitude to be well aware of the broad factual background from which the framework of this workshop construction emerged. To those who never dare to look into microscopes or at the sky, who never sit in an orgone energy accumulator and yet are full of fake "authoritative" opinions about orgonomy, we say in advance: step aside and do not disturb most serious work. Keep quiet, at least!

Years of painstaking observations and functional theory formation have hewn two major pathways into the realm of non-living nature that revealed the function of superimposition to be at work at the very roots of the universe. One pathway leads into the microcosmos, the other into the macrocosmos. Superimposition is the CFP that integrates both into one natural function.

Let us begin with the microcosmic realm. We shall not dwell too long in it since, though the theoretical outlines seem clearly marked, there are many gaps in details essential to a firm foothold. The essence of the microcosmic framework is as follows:

In completely darkened, metal-lined orgone energy observation rooms we can observe luminating orgone energy units
pursuing certain pathways as they move spinning forward
through space. These pathways distinctly show the form of a
spinning wave.

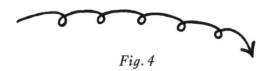

Fig. 4

This was reported on several occasions many years ago
without further elaboration. There is now ample, well-
reasoned evidence to the effect that *two such spiraling and
excited orgone energy units attract and approach each other
until they superimpose.* Thus:

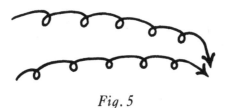

Fig. 5

It is an essential characteristic of our base of operation to
assume that the primordial orgone energy ocean is entirely
mass-free. Accordingly, mass (inert mass at first) emerges
from this mass-free energy substratum. It seems logical further to assume that *in the process of superimposition of two
mass-free, spiraling, and highly excited orgone energy units,
kinetic energy is being lost, the rate of spiraling motion de-
creases greatly, the path of motion is sharply curved, and a*

*change takes place from long-drawn-out spinning forward
toward circular motion on the spot.*

Exactly at this point of the process, inert mass emerges
from the slowed-down motion of two or more superimposed
orgone energy units. It is immaterial whether we call this
first bit of inert mass "atom" or "electron" or something
else. The basic point is the emergence of inert mass from
frozen kinetic energy. This assumption is in full agreement
with well-known laws of classical physics. It is also in agreement, as will be shown in a different context, with the quantum theory.

To continue our train of thought, we must further assume
that the material, chemical "particles" that compose the atmosphere have originally emerged and are still continuously
emerging through superimposition of two or more spinning
orgone energy units in the orgone envelope of the planet. It
matters little at this point in what particular manner the
different material units are created from primordial orgone
energy. We restrict our curiosity to the above-mentioned
basic change:

INERT MASS IS BEING CREATED BY SUPERIMPOSITION OF
TWO OR MORE SPINNING, SPIRALING ORGONE ENERGY UNITS
THROUGH LOSS OF KINETIC ENERGY AND SHARP BENDING
OF THE ELONGATED PATH TOWARD CIRCULAR MOTION.

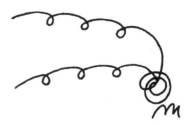

*Fig. 6. Creation of the primordial mass particle (m)
through orgonotic superimposition*

A functional relationship is hereby established between the spinning movement of mass-free orgone energy (OR) and inert mass (m), which also characterizes the relationship of heavenly bodies spinning in the surrounding orgone ocean. Spheres or discs of solid matter spin on a spiraling path within a faster-moving, wavy orgone energy ocean, as balls roll forward on a faster-moving, progressing water wave. The exact numerical relationship of the two movements, though of great importance, does not matter at this point. What is important is that a functional relationship has been found between the movements of primordial orgone energy and matter that, for the first time in the history of astrophysics, makes comprehensible the fact that heavenly bodies move in a spinning manner. Furthermore, it makes comprehensible the fact that our sun and our planets move in the same plane and in the same direction, held together in space as a cohesive group of spinning bodies. The spinning wave is the integration of the circular and forward motion of the planets, of their simultaneous rotation on the N–S axis and their movement forward in space. *The orgone ocean appears as the primordial mover of the heavenly bodies.*

Sharply delineated, new astrophysical problems arise that cannot and should not be discussed at this time. It is sufficient to have them tentatively formulated:

1. It is necessary to assume that the first material particles that were "created" by superimposition of two or more orgone energy units form the nucleus for the growth of the material body. It does not matter at present whether these "core" elements of the future heavenly body are of a gaseous or of a solid nature, or whether they possibly go through a process of development from a gaseous to a solid state. What matters is that a starting point for the development of a heavenly body from primordial energy has been hypothetically established. (Cf. Bibliography, No. 31.)

2. A further logical necessity is the assumption of a GENESIS *of the function of gravitational attraction.* The growth of the material core particle of the future heavenly body would be accomplished on the basis of the *orgonomic potential.* The orgonotically stronger body attracts smaller and weaker systems, such as mass-free orgone energy units, and other small bits of primordial matter as they arise in the orgone ocean, which surrounds the first growing core. It would be necessary, furthermore, to distinguish between the *orgonotic attraction* of two energy waves and the *gravitational attraction* between two material bodies, i.e., to establish that primordial orgonotic attraction changes functionally into gravitational mass-attraction.

3. From points 1 and 2, we would further have to assume that the growing material core would be permanently surrounded by an orgone energy field which from now on would be subject to the gravitational attraction of that core. This would explain the origin of the orgone envelope of the sun (corona) and of the earth. Both are clearly visible and are governed by basic orgonomic functions such as wavy motion from west to east, faster motion of the envelope than the globe lumination, blue color, and containment within the field of attraction of the material core.

4. The mass-free orgone energy stream that surrounds the material globe, must, due to the orgonotic attraction exerted upon it by the core, separate from the general stream of the cosmic orgone energy ocean and follow the rotation on its axis of the material body. Thus, *the cosmic ocean, hitherto unitary, splits up into one major and one minor orgone energy stream.* This assumption will be verified by concrete astrophysical functions (cf. fig. 7).

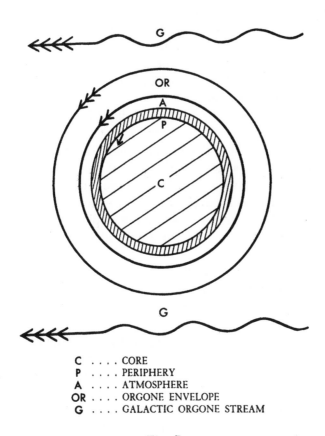

C . . . . CORE
P . . . . PERIPHERY
A . . . . ATMOSPHERE
OR . . . . ORGONE ENVELOPE
G . . . . GALACTIC ORGONE STREAM

Fig. 7

5. The gaseous atmosphere that surrounds the heavenly bodies would necessarily have to emerge through superimposition of mass-free orgone energy units in the revolving orgone energy envelope. This necessary assumption would have to be confirmed in due time by the establishment of the laws that lead from the mass-free orgone energy units to

the atomic weights of the gas particles that constitute the gaseous atmosphere.

6. It follows that concentration and condensation would increase toward the core of the rotating body, the heavier elements being located near the center and the lighter elements progressively nearer the periphery, with the lightest gases—helium, hydrogen, argon, neon, etc.—being located at the extreme periphery.

7. In this connection, a most striking functional identity must be mentioned, which so far has not attracted attention in scientific thinking. The chemical elements that constitute the gaseous atmosphere of the planets are identical to the elements that constitute the living orgonotic systems. They are: hydrogen (H), oxygen (O), nitrogen (N), and carbon (C), and their various molecular groupings such as CO_2, H_2O, $C_6H_{12}O_6$, etc. This functional identity must have a deep significance.

The functional identity concerns only the primordial orgone energy functions and the transformations from primordial mass-free to secondary mass-containing functions. From there on, but *not* previously, the well-known laws of mechanics and chemistry are fully valid. They submit to evolution; they have a genesis. The problem to be solved in detail is HOW THE MECHANICAL AND CHEMICAL LAWS ORIGINATED FROM THE FUNCTIONAL PROCESSES IN THE MASS-FREE PRIMORDIAL ORGONE ENERGY OCEAN.

The advantage of our work-hypothesis, as delineated above, is quite obvious. To summarize:

1. It frees us from the clumsy assumption of material bodies rolling in an "empty space," in a merely mathematically approachable action at a distance in a "field." The "field" is *real*, of a *measurable, observable,* and thus *physical* nature. *Space is not empty* but is filled in a continuous manner without gaps.

2. It frees us, furthermore, from the uncomfortable idea

that a gravitational attraction, which never could be demonstrated, is exerted by the sun over tremendous distances upon all the planets. **The sun and the planets move in the same plane and revolve in the same direction** *due to the movement and direction of the cosmic orgone energy stream in the galaxy.* **Thus, the sun does not "attract" anything at all. It is merely the biggest brother of the whole group.**

We have done no more than draw a sketch of the transition from the microcosmic to the macrocosmic function. We shall later return to superimposition in the macrocosmic realm in greater detail. But first we must acquaint ourselves with some important functions pertaining to the function of superimposition in the living realm, where it was originally discovered.

We shall concentrate upon two basic functions only:

1. *The spinning flow of orgone energy in the living organism* ("bio-energy").

2. *The superimposition of two orgone energy streams in living bodies,* COPULATION, and the functional meaning of the drive to genital embrace and orgastic discharge.

CHAPTER IV

THE LIVING ORGONOME

The formation of living matter in orgone Experiment
XX[1] combines numerous bio-energetic and biofunctional phenomena into a single result of great significance. This experiment reproduces the process of *primary biogenesis*, i.e., the
original formation of plasmatic, living matter through condensation of mass-free cosmic orgone energy. This conclusion derives logically from the fact that organic forms with
all the properties of living matter (*structure, pulsation, reproduction, growth, and development*) can be developed by
a freezing process in a clear solution of high orgonotic potency. The subject is inexhaustible, but it is not our objective to treat it exhaustively. Once more, I would recall the
discovery of the American continent by Columbus. This discovery did not exhaust all past and future possibilities of
America. It did open the door to an enormous territory full
of future potentialities. The same holds true for Experiment
XX.

The schema below represents approaches to the manifold
functions of nature that were opened up by Experiment
XX:

---

[1] Cf. *The Cancer Biopathv*

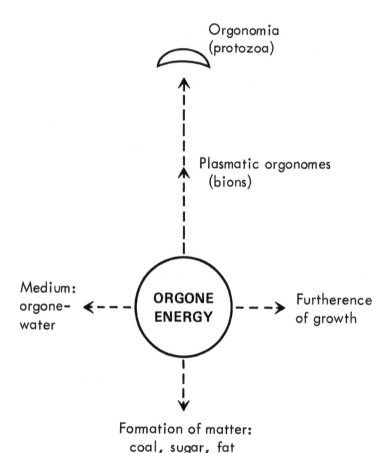

1. The development of organic forms, plasmatic "orgonomes" (bions).
2. The organization of protozoa (orgonomia).
3. The formation of biochemical matter: coal, sugar, fat.
4. The furtherance of life and growth as affected by the orgone-water solution.

In the process of the freezing experiment, energy is transformed into matter. This matter is alive. By means of dehydration or burning of the flakes, carbon and a sweet-tasting, sugary substance originate from it. These are gross characteristics to be elaborated in detail. In this process, frozen orgone energy passes through all phases of bionous formation revealed by orgone biophysics: T-forms develop into PA bions through the intake of mass-free orgone energy; the PA bions grow into larger, rounded shapes resembling small "eggs"; some of these "egg shapes" expand and become bean-shaped; the bean shapes acquire motility and form protozoa: ORGONOMIA. In movement and shape, they look very much like spermatozoa. We may assume that the spermatozoa and eggs in the metazoa are also formed through condensation of orgone energy in the germinal tissues. The development of formed bions from distilled orgone water establishes beyond any doubt the process of *primary* formation of organic matter from mass-free orgone.

Bion water is yellow, ranging in intensity to almost a brown. In this context, one is reminded of the yellow resin produced by trees, of yellow honey produced by bees, of the yellow color of animal blood serum, the yellow of urine, etc. Also of great significance is the "blood sugar level" in the organism. Thus the gap in biology, which up to now contained a mystery—namely, how plants convert "solar energy" into carbohydrates and solid cellulose forms—is apparently closed. "Solar energy" is our orgone energy that the plants absorb directly from the soil, the atmosphere, and the rays of the sun.

The leaves of evergreen ivy are a case in point. In winter, the leaves lose their green color, except for the venation, which remains green. The rest turns a yellow-brown. In spring, the green expands from the leaf vessels across the smooth leaf. This phenomenon permits the assumption that, in winter orgone energy retreats from the periphery of the leaves; in other words, it contracts because of the cold, to expand again in spring. That portion of the ivy leaves about to die off is thus revived.

The change from green to yellow in autumn and from yellow to green in spring becomes perfectly comprehensible in terms of orgonotic functioning. According to classical investigations, green is the result of a mixture of yellow and blue. Blue is the specific color of orgone energy, visible in the atmosphere, ocean, thunderclouds, "red" blood cells, protozoa, etc., and on orthochromatic photographic plates after irradiation with earth bions.

Now it seems clear that the yellowing of the leaves in autumn is due to disappearance of the blue from the green, and accordingly, the turning toward green again in evergreen ivy is due to new absorption of orgone energy from the atmosphere. Thus, the green of leaves is the result of the mixture of yellow resin and blue atmospheric orgone energy.

At present, we would like to limit our investigation to a single function: the origin of formed living matter from mass-free orgone energy. Right now we are not interested in the chemical composition of these forms.

There is only *one* assumption that satisfactorily explains the origin of motile, formed living substance in Experiment XX. *In the process of freezing, the mass-free orgone energy in the fluid contracts, just like living plasma.* Hence, this contraction does not depend on the existence of formed matter. It exists *prior* to the formation of matter, as a basic function of cosmic orgone. *The contraction of orgone energy is accompanied by condensation, and condensation is accompanied by the formation of material particles of microscopically small dimension.* The classical, mechanistic concept does not provide for any causal connection between energy *movement* and organismic *form*. *Orgone biophysics*

*can prove a functional connection between form of move-
ment and form of living matter.*

Primary matter originated in the cosmos, and the process
of matter formation apparently continues uninterruptedly.
The cosmic origin of bio-energy is experienced as an equation of "life-earth-sun-spring." The mechanistic concept
knows only atoms and molecules that combine to form salts
and organic bodies. It can explain neither movement nor formation of living matter because neither the first nor the second resemble mechanical movements and known geometric
forms in any way. *In contrast, orgone biophysics operates
with a concrete cosmic energy. It postulates that the functions of cosmic energy in the realm of inorganic matter are in
harmony with those in the realm of living matter.*

In Experiment XX, membranes and then bions are formed
from mass-free orgone energy. They constitute forms that
cannot yet be described as "living organisms" in the accepted biological sense, but they already show the typical
shape of living organisms. This is clearly apparent in the
illustrations of Experiment XX (cf. fig. 19). The forms of
most flakes resemble those of fish or tadpoles. Now, if forms
invariably express frozen movement, we may reason, a posteriori, from these forms to the functions of orgone energy.
Exact observation and extensive comparison will show that
there exists a *basic form of living matter* that has no counterpart in classic geometry. Viewed laterally, this basic form
looks as follows:

I. SIDE VIEW:

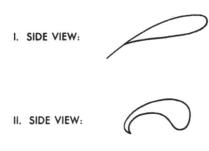

II. SIDE VIEW:

Fig. 8

Viewed from above or below, the living form is typically as follows:

III. FRONT VIEW:

Fig. 9

Before investigating the bio-energetic function of this form, let us make sure that it is indeed *the* biophysical basic form. It clearly applies to:

1. Plant seeds: wheat, corn, barley, oats, beans, lentils.
2. Plant bulbs: potato tubers, almond kernels; the pits of apples, pears, plums, peaches.
3. Animal sperm cells.
4. Animal eggs, particularly birds' eggs.
5. Animal embryos.
6. All organs of the animal body: heart, bladder, liver, kidney, spleen, lung, brain, testicle, ovary, uterus, stomach.

7. Unicellular organisms: paramecia, colpidia, vorticellae, cancer cells, human vaginal protozoa (trichomonas vaginalis) etc.

8. Whole animal and plant bodies: jellyfish, starfish, reptiles of all kinds; the trunk formation of all kinds of birds, fish, beetles; mammals, including man, etc.

9. Trees in general, as well as each single leaf and blossom; pollen and pistils of plants.

It is noteworthy that even those organs extending from the trunk—arms, legs, fins, wings, the head of the snake, the lizard, the fox, man himself, the fish, etc.—in turn take the form of the "orgonome." Even the claws and beaks of birds, the air bladder of fish, the horns of cattle, rams and stags, the shells of snails and mussels take the form of the "orgonome."

All this points to the work of a natural law of energy, a law that fundamentally differs from the geometric laws of classic mechanistics.

Access to this law of cosmic energy must be sought in the movement of mass-free orgone energy.

Just as the *expressive movements* of living matter are inextricably tied to an *emotional expression* that is meaningful in relation to the world around it, so the form of living matter has a specific expression too. The point is to read it correctly.

All forms in the realm of living matter can easily be reduced to the egg form without violating the individual variations of form. This basic form varies with length, width and thickness. It may appear in subdivisions of the same form, as in worms; but whether as a whole or in part, the basic form of living matter always remains the same egg form.

Such a consistent uniformity of the organic form must correspond to a fundamental law of nature and a natural law of cosmic dimensions. For the basic biological form is universal, regardless of climate or geology. It is as if cosmic orgone energy, in organizing living substance, obeyed only one law, its own law of motion.

We shall call the specific basic form of living matter the ORGONOME. Its typical basic form is the following generalization of microscopic forms from Experiment XX:

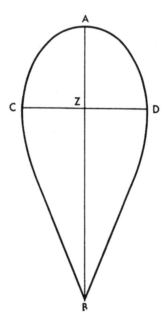

Fig. 10. Closed orgonome, basic form

TRIGONOMETRY OF THE ORGONOME

We would like to designate as an orgonome that specific form which, in its purest shape, is represented by the hen egg.

The orgonome is neither a triangle nor a square nor a circle; it is neither an ellipse nor a parabola nor a hyperbola. The orgonome represents a special, novel geometric figure, a closed plane curve, not unlike the ellipse with half-axes of varying length and width, but differing from the ellipse precisely because of the different length of the large axes.

Let us try to determine how an orgonome originates, orgonometrically speaking. Two fundamental natural phenomena are involved:

1. The orgastic convulsion.
2. The spinning wave—Kreiselwelle (KRW).

We encounter the orgastic convulsion in the entire animal kingdom. We discern the spinning wave (KRW) by observing atmospheric orgone in the darkroom. The tiny, blue-violet points glide along specific trajectories, which I described schematically in the second volume of *The Discovery of the Orgone* as follows:

Fig. 11

Let us isolate an individual wave from the KRW path:

Fig. 12

If we place two such spinning waves together with their concave sides, we obtain the known form of the ellipse:

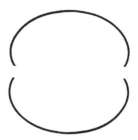

Fig. 13

However, if we bend a KRW in the center at point A and bring the two ends of the KRW—B and B'—together, we obtain the egg, or orgonome form.

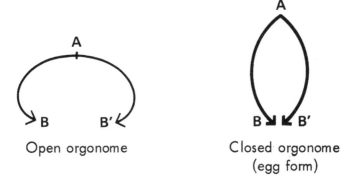

Open orgonome   Closed orgonome
                                           (egg form)

Figs. 14 and 15

We could work out this process in terms of pure trigonometry without providing orgone-physical proof. But the orgastic convulsion gives us a *biophysical* argument that endows this trigonometric process with great importance. The most conspicuous phenomenon in the orgasm reflex is the *striving of both ends of the torso—the mouth and the genitals—to come closer together.* This biophysical phenomenon indeed put me on the trail leading to the origin of the orgonome form. In the orgastic convulsion of an animal, or in the swimming motion of the jellyfish, the body seems to sag in the center, bringing both ends closer together.

The connection of a fundamental *biological* movement with a *physical* movement may, at first glance, seem arbitrary. But such a connection is justified if it opens the door to an obvious lawfulness in biological functioning. To my knowledge, the basic form of living bodies has never been understood. And if the orgasm reflex promises us an understanding of the orgonome form, we must not reject it.

The similarity of a KRW to an animal body, viewed laterally, is indeed startling (cf. fig. 14). Detailed proof of this similarity cannot be presented here but has already been established experimentally.

If living matter is frozen orgone energy, the form of movement of the orgone energy must necessarily translate itself into the form of living matter, the orgonome form. This functional continuity is hard to find in the realm of inorganic matter. It is easily understandable in the realm of living matter. *If form is the movement of energy that is frozen, then the organ form must derive from the form of movement of cosmic energy.*

Let us return to the orgasm reflex, this rich source of bioenergetic insights:

We found that the orgasm reflex cannot be verbalized in terms of idiomatic language. Its mode of expression, we concluded, was supraindividual—neither metaphysical nor mystical, but cosmic. In the orgasm reflex, the highly excited organism attempts to bring both ends of its torso closer together as if to unite them. If this interpretation is correct, it must also prove correct in other categories of orgone functioning and cannot be limited to the orgasm reflex alone.

Let us now look at the form of the biological orgonome in its functional connection with the form of plasmatic currents. True to the principle of the functional identity of all living substance, we must gather apparently widely separated functions and look for their common denominator.

Plasmatic current does not flow continuously but in rhythmic thrusts. Hence we speak of PULSATION. The pulsation can be plainly observed in the blood circulation of all metazoa. The pulsatory current of body fluids is the work of the organismic orgone, a direct expression of its form of movement. From the pulsation of body fluids we must reason, a posteriori, that there is a pulsation of orgone energy. This conclusion is confirmed by observing certain protozoa, in which pulsatory waves of excitation pass through the body and set the protoplasm in motion. Among worms, excitation waves of a pulsatory nature pass from the tail end to the head. The same phenomenon can be seen in certain amoeboid cancer cells. The following drawing expresses the form in which excitation waves move in the protoplasm of these cancer cells:

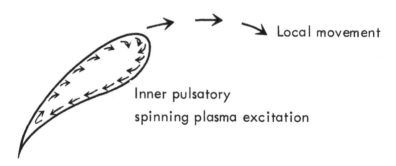

Fig. 16

Hence we must distinguish between two kinds of pulsatory movements in living matter: *the pulsatory movement of orgone energy* in the organism, and its effect, the *pulsatory mechanical movement of body fluids.* We differentiate here between functional bio-energetic pulsation and mechanical pulsation. *The mechanical pulsation results from the functional pulsation of the orgone,* its spinning forward in alternating expansion and contraction.

Since the movement of fluids is mechanical, it can only be the expression and consequence of the pulsatory function of orgone energy. Among the flowing amoebae, the bio-energetic pulsation coincides completely with the organic flow of fluids. Among the colpidia and paramecia, the body is rigid and contains large, membranous fluid-filled vesicles without flowing plasm. Here the movement of energy can be discerned only in the locomotion of the whole body. If we compare the form of movement of the waves of excitation in cancer cells with the external form of movement of trichomonas vaginalis, colpidia, and paramecia, we find there is a thrusting, pulsatory motion that does not proceed in a straight line but in the manner of a spiral, presenting an overall curvature. We can connect the individual points of the movement curve and find a geometric figure that depicts a spinning wave (KRW) and looks roughly as follows:

Fig. 17

We see that the curve of the plasma current inside the body of the cancer cell is the same as in the locomotion of the whole body of a colpidium. If we dissect the curve of the orgonotic plasma current into its individual parts, we obtain a shape that, laterally viewed, resembles the form of all living organs and organisms (cf. fig. 14).

This harmony, in the form of movement of the energy particles, plasma current, orgonotic excitation waves and the shape of the organs, cannot be mere coincidence. It is obviously governed by a common law of movement revealed time and again in the individual forms of motions and structures. Even the elongated earthworm, which, at first glance, reveals nothing that resembles an orgonome form turning back on itself, shows the orgonome in the segments. Furthermore, the earthworm curls up in a manner that looks like the orgonome of a snail shell (cf. fig. 20 : 3 and 4).

The following diagram illustrates the structuralized, clearly expressed original movement of the organismic orgone energy in the growth of a shell:

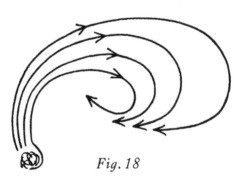

Fig. 18

Thus we can distinguish three states of orgonotic expressive movement:

a. *The spinning motion of orgonotic excitation waves, of protoplasm and of the locomotion of protozoa.*

THE LIVING ORGONOME 207

b. *The orgonome form of animal organs and organisms, i.e., frozen orgone movement.*
c. *The orgonome form of the animal body at rest, as an intermediary state between energy movement and solid matter.*

We now have a better biophysical understanding of the segmentary arrangement of the orgonotic current in man and of the segmentary arrangement, or armoring, in the biopathic character.

The plasmatic (mechanical) and the orgonotic (bio-energetic) currents in man—blood circulation and excitation waves—have the same rhythmic, wavy, and segmentary character as observed in the earthworm. *The segmental arrangement of the armoring expresses the immobilization of individual parts of the wave path,* or, to put it differently, one wave freezes into one formed orgonome segment.

Thus the principle of orgone therapy—to proceed always from the "head" to the "tail," i.e., to the genitals—acquires its bio-energetic meaning. As in the earthworm, the snake and the plasmatic cancer cell, the orgonotic waves invariably pass from the tail end over the back toward the head. Bio-energetically, this arrangement of the orgonotic flow makes sense, because it is predicated on the "forward" movement of the whole body in the direction of the head. In orgone therapy, if we first loosened the armoring at the tail end, the liberated energy would be blocked at the segment located farther ahead. But the dissolving of the armor at the head end eliminates the armor rings at the place *toward which* the orgonotic excitation must flow. We meet the direction of the current, and thus free the way for its unhindered flow, instead of starting to break the armor at the source of this current. The technique of orgone therapy did not start out with these biophysical speculations in mind, but followed purely clinical considerations, e.g., that it would be advan-

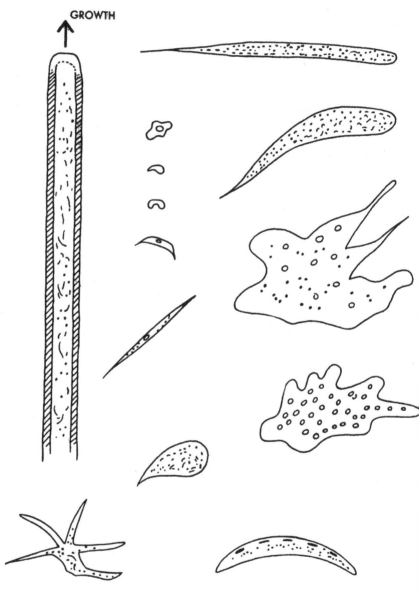

GROWTH

Fig. 19. *Various typical forms of plasmatic flakes in Experiment XX, drawn from nature: bio-energetic orgonome*

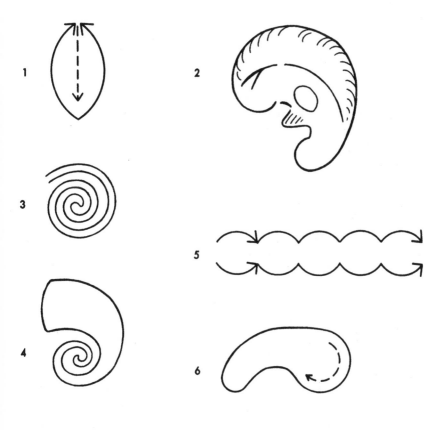

1. Two extended, open orgonomes, placed together, show the heart form; also the form of tree and plant leaves, various fruits (plums, etc.), eggs
2. Human earlobe, shells of oysters, clams
3. Curled-up worms, snakes
4. Shell of the snail
5. Intestines, worms, caterpillars
6. Embryo, stomach, brain, spleen, kidney, liver, pancreas

Fig. 20. Various orgonome forms, abstracted

tageous to liberate all the energy of the body before mobilizing the genitals. But, as we now see, the clinical and the bio-energetic aspects of the matter combine in a common useful function.

Let us now return to our Experiment XX in order to learn more about the formation of living substance into the orgonome. We find plasmatic flakes in which first circular, then bean-shaped orgonome forms can be seen. In the bean shape, the orgonome is once more clearly evident. This orgonome is in motion. Its movements again have orgonome form, as can easily be discerned in the spiral lines of their progressive movement.

We may now appropriately conclude that *through the freezing process the freely moving orgone energy in the fluid is, in very small part, converted into matter by membrane formation.* Since the movement of orgone energy is curved, it stands to reason that the membranes are also curved. Inside the membranes, mass-free orgone energy continues to move. Naturally enough, it strives to expand the membrane, as if it meant to burst through the sac in which it is trapped. There is of course no reasoning involved here, but rather a contradiction between the function of the expanding movement of the mass-free orgone and the confining membrane. Logical deduction demonstrates that nothing but a bean shape, our orgonome, can result from this contradiction between energy flow and restricting membrane.

Of course, the formation of the bean shape does not in any way satisfy the motile impulse of mass-free orgone energy inside, an impulse directed toward *stretching the curve,* i.e., toward moving away from the spot. Therefore, the local forward movement, whose basic tendency consists again in stretching, curving, and rhythmically reverting upon itself, appears for the first time.

The development of colpidia from primary embryonic vesicles is particularly suited for studying the plasmatic currents that are set in motion by the orgone energy in the membranous sac. As soon as a membrane has formed around a cluster of bions, the budding germinal vesicle appears. The interior shows a vesicular structure and a blue glimmer. The membrane is taut, but the whole system is still at rest (fig. 21 : 1). That motile impulses are freed in the interior of the "germinal vesicle" is shown by a rolling motion of the vesicles occurring sooner or later. While the membrane rests, the vesicles at first roll near the periphery, in *one* direction along the membrane. The inner cohesion loosens. Along with the rolling motion in one direction goes a reciprocal attraction and repulsion. After a while, the direction of the movement changes; the vesicular content reverses its direction. In this manner, the bionous content gains elasticity (21 : 2). The germinal vesicle tautens more and more; it grows larger. *Gradually the circular form turns into the egg form, our orgonome form.* The plasmatic current at one end splits into two currents. The two currents converge and continue backward along the center line (21 : 3). Now we can clearly distinguish *two* halves of the orgonome, each of which assumes more and more clearly the bean shape, or lateral orgonome form. After several hours of strong orgonotic motility of the plasm, the germinal vesicle usually bursts into *four* "complete" colpidia. So far, we cannot determine whether the figure "four" is the rule or whether a division into two colpidia also occurs. What is important is that *the forward end of the colpidium is located at the place where the current was originally directed. The animalcule swims off locally in the direction of the original plasma current* (21:4). This current has assumed an orgonome form. Now, when the local movement begins, the internal current stops and the animal moves forward as a whole in lines that are slightly curved. *The curve of the path of locomotion is identical with the curve of the "back,"* as illustrated in drawings from life (fig. 21).

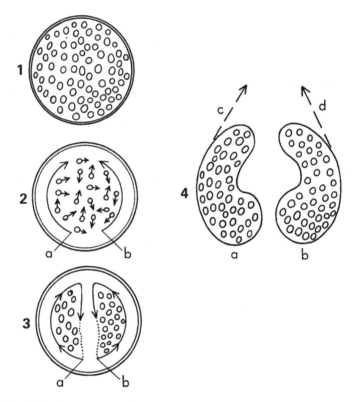

Fig. 21. Schematic illustration of the development of the closed orgonome into the open orgonome in the colpidium

1. Germinal vesicle of the colpidium, at rest.
2. Germinal vesicle, with internal motion: a and b illustrate the alternating directions of the rotating motion of the tiny energy vesicles.
3. Division of the orgonotic excitation waves; beginning of two orgonomes. Arrows point to the converging currents.
4. The two closed orgonomes a and b move forward in space into the open orgonomes c and d.

Let us summarize the processes in the living orgonome.

1. The inner motility is nourished by wave-like pulsating orgone energy that is trapped in a membranous "sac."

2. The movement of the orgone energy is responsible for the inner motility of the structured bionous substance.

3. Since the inner orgone movement is confined by the membrane, a curved path of the plasma current is produced, in which we recognize the orgonome.

4. The "energetic" orgonome leads to the formation of the material orgonome. The form of the organs reflects the form of the original energy movement.

5. There is a contradiction between the movement of the orgone energy and the taut membrane. The membrane sharply deflects the original forward movement of the current backward. Since this happens at all the curvatures of the vesicle, the currents converge toward the center and thus produce a division of the vesicle into four structural orgonomes.

6. Once this division is complete, we observe the separation and local forward movement of the individual orgonomes. The local movement proceeds in a curved line—a motion with alternating long and short half-waves. The motion "away from the spot" is obviously dictated by the direction of the orgonotic impulses. It is curved in terms of the "back." The fore end is always located in the direction of the original orgonotic current.

ORGONOTIC SUPERIMPOSITION

To summarize: The specific orgonome form of living matter and its organs results from an opposition between mass-free orgone energy and frozen orgone that has become membranous matter. Mass-free orgone always strives to break beyond the enclosure of the membrane. *The bio-*

energetic orgonome is extended and open; the material orgonome is closed. Since the excitation waves of the bio-energetic orgonome move within the limits of the closed material orgonome, they necessarily press against the membranous boundary, as shown in the following drawing:

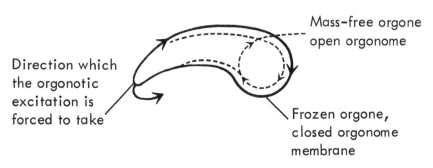

Fig. 22

This creates a *stretching* of the orgonome, in which we recognize the basis for all kinds of growth, particularly as shown in the stretching of the gastrula as it becomes the typical elongated embryo of a multicellular organism (metazoan).

The function of growth corresponds to the expansion of the membranes of the closed orgonome. That this indeed involves expansive functions of mass-free orgone energy can be seen from the curved protrusions that initiate the formation of every new organ in the embryo of all animal species. Again, the embryonic protrusions show the typical form of the orgonome.

The elasticity of the formed body membrane and the presence or absence of a skeleton determine how much of the original spinning motion of the bio-energetic orgonome is clearly evident. But even where a fully developed skeleton and an extensive muscular structure have blotted out the external appearance of the excitation waves, there is still the rhythmic excitation and current pulse of the blood circulation, as well as the orgonotic current or plasma excitation, which are felt subjectively. In the orgasm reflex the original form of movement of the bio-energetic orgonome is unmistakably perceived insofar as it seizes the entire organism.

We distinguish the following kinds of SUPERIMPOSITION:

1. Two *open* bio-energetic orgonomes place themselves on top of each other

2. Two *open* orgonomes place themselves side by side

3. Two *closed* orgonomes are superimposed upon each other

Fig. 23

The superimposition of two closed orgonomes is the bio-energetic basis for the superimposition of two organisms during copulation (cf. fig. 25). In this process, the highly excited tail ends penetrate each other bodily; the two orgonomes merge bio-energetically to form a single highly charged energy system. It is characteristic for the homogeneousness of all processes in the living realm that the energy functions of excitation, superimposition, interpenetration, and fusion are repeated in the same functions of the reproductive cells. For, during copulation, sperm cell and egg cell continue the function of superimposition and fusion of the male and the female orgonome, although the division of living orgonomes into male and female individuals remains mysterious even from the standpoint of orgone physics.

Let us now try to comprehend the expressive movement of the orgasm reflex on the basis of the orgonome as the fundamental biophysical form of living matter.

It cannot be the function of the orgasm reflex, as one might assume from the purely teleological standpoint, to carry the male semen into the female genital organ. The orgasm reflex occurs independently of the ejaculation of semen, because we also find it in the embryo—in the typical forward position and convulsion of the tail end; in the rocking, bio-energetic forward motion of the tail end of many insects, such as wasps, bees, and bumblebees, as well as in the usual position of the pelvis and the hind legs among dogs, cats, and hoofed animals. These examples should suffice to demonstrate that the orgasm reflex has a far more general life function than mere fertilization. The mechanistic and finalistic interpretations do not work in this area; they are too narrow and do not reach the heart of the matter.

Let us try to interpret the function of the orgasm reflex in terms of its expressive movement.

The living orgonome, be it an embryo, an insect, or a more

highly organized animal, is essentially characterized by the following:

First, local forward motion invariably and logically proceeds in the direction of the larger and wider fore end. Second, the genital organs are invariably and logically located on the ventral side near the tail end. Third, in a state of orgonotic excitation of the orgonome, the genital organ expands through erection in the direction of the local forward motion. Fourth, the movements that cause the interpenetration and fusion of the male and the female genital organs drive the entire tail end to the fore in a highly energetic manner (cf. fig. 24).

These biological phenomena are valid for the animal kingdom at large, except for those species that have barely progressed beyond the stage of the primitive orgonome form of the jellyfish. Although they seem to be far apart, there is nevertheless a close functional interconnection. It can be found if we are again guided by the process of orgonotic excitation.

The form and position of vertebrae among vertebrate animals reveal the direction of orgonotic excitation waves during growth: they always start from the tail end and move over the length of the curved back toward the head end. They also follow the same direction during the entire lifetime of the organism. This can be experienced subjectively if shudders of pleasure or fear pass over one's back. The fur of frightened animals can be seen to "stand on end" due to contraction of the mm. erectores pilorum in the direction of the orgonotic wave motion, leaning *forward*.

As we can see from the drawing (fig. 24), the entire back is gently curved and as such is in harmony with the curved path of the orgonotic waves. Presumably, the curve of the wave path conditions the curve of the back, and not the other way around. But once the material, closed orgonome is

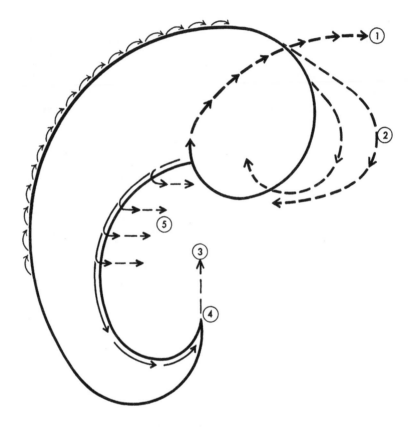

1. Direction of *forward movement*: antennae, optic peduncles, primary brain vesicles
2. Direction of growth
3. Org-movement continuation of the direction
4. *Greatest orgonotic excitation*, sharpest deflection
5. Intermediary outbreaks

Fig. 24. Direction and results of movement within the closed orgonome (orgasm reflex)

formed, it confines the bio-energetic waves of excitation and forces them to deflect from the original path of the extended course. It is probable that the generally frontward formation of the secondary protrusions during the growth of the embryo is associated with this process. Here, the essential point is the opposition between the *material* and the *bio-energetic* orgonome. The membrane of the material orgonome returns from the fore end to the tail end, forming a characteristically wide curvature. In the animal embryo the curvature of the orgonome turns in at the neck toward the body center, then turns away from it near the chest. The curvature of the orgonome forces the excitation waves back toward the tail end. Part of the orgonotic excitation apparently is indeed deflected toward the tail end, but another part pushes through the membrane at the fore end in the direction of the original excitation waves of the bio-energetic orgonome.

As long as the directions of the material and the bio-energetic orgonome are in harmony, there are no new formations and no directions of movement of the whole. The body orgone does not press outward from the orgonome sac. Therefore, no organs are formed along the length of the backs of animals, no protrusions of any kind, but neither is there any movement in the direction of the back, nor any growth. The humps on the back of the camel or the dorsal fins of certain fish are exceptions that remain to be explained.

Growth in the vertical (longitudinal) *axis and local forward movement thus appear as functions of the body orgone energy, the result of its attempt to burst through the confining membrane sac.* The membrane "goes along," i.e., it expands and thus forms the protruding sacs of the organs in their primitive condition.

In contrast to the back, where the material and bio-energetic orgonomes are in harmony, we find at the fore end on the ventral side a multitude of organ formations of various kinds: the domed forehead, the nose or snout, chin, breasts, the limbs, and the genitals. Now, if our functional concept of organ formation is generally valid, organs formed by protrusion of membranes must always originate at the ventral side, where the direction of the current of biological energy is deflected from its regular course, i.e., wherever the body orgone "strives to break out of the sac" (cf. fig. 24).

We see from our drawing that the progress of the membrane on the ventral side indeed runs counter to the original and true direction of the orgone waves. Consequently, we find time and again, at almost regular intervals—such as in the arrangement of the limbs and the nipples of the breasts —a rhythmically recurring tendency to break through. This contradiction between membrane and energy wave reaches its culmination at the tail end. The tail end is pointed and sharp; the material orgonome moves sharply forward again in the direction of the forward movement of the waves of excitation.

The strong forward propulsion of the tail end among animals, based on concentrated orgonotic wave excitation pressing outward, explains the "genital excitation" and the orgasm reflex in a satisfactory and probably complete manner. The overwhelming pressure of the orgonotic excitation in the pointed and narrower tail end and especially in the less spacious genital organs is explained by the concentration of orgone waves in a narrow space. *The orgone energy, deflected from the head to the tail, i.e., opposite its natural direction, presses toward the genital organ in the original forward direction, exciting it and forcing it forward into erection.*

We can now interpret the copulation of animals from a functional bio-energetic, i.e., orgonomic, standpoint. The orgone, pressing forward and concentrated in the genital organ, cannot escape from the membrane. *There is only* ONE

*possibility of flowing out in the intended direction—fusion
with a second organism, in such a way that the direction of
excitation of the second organism becomes identical with the
direction of the orgone waves in the first.* This process is
actually achieved in orgonotic superimposition as shown by
the drawing (fig. 25). We see that, with the superimposition
of the two orgonomes and with the interpenetration of the
genitals, the pressed and therefore "frustrated" tail end can
allow its orgonotic waves of excitation to flow in the natural
direction, without having to force them back sharply, and
that furthermore the space in which these waves can run
their course is widened.

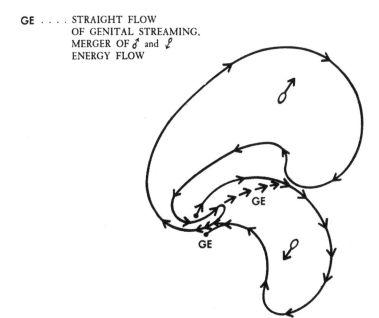

GE . . . . STRAIGHT FLOW
OF GENITAL STREAMING.
MERGER OF ♂ and ♀
ENERGY FLOW

*Fig. 25. Function of "gratification" in genital
superimposition*

Our assertion, according to which the orgasm reflex has no immediate linguistic meaning, is correct. Its function lies beyond language. Yet it expresses something concrete: superimposition follows orgonotic interpenetration. *The preorgastic body movements and especially the orgastic convulsions represent extreme attempts of the mass-free orgone of both organisms to fuse with each other, to reach into each other.*

I said earlier that the bio-energetic orgonome always strives beyond the realm of the material orgonome. *While the energy of one organism flows into the energy system of the second organism, mass-free orgone energy actually succeeds in transcending the limits of the material orgonome, i.e., the organism, and, by merging with an orgonotic system outside its own, it continues to flow.* This takes into account the tendency toward stretching, toward widening the effectual area of mass-free orgone energy. In the acme of excitation, large quantities of energy are indeed flowing out, along with genital substances. This process is connected with the subjective sensation of "release," "liberation," or "satisfaction" ("gratification"). Since language directly reflects the function of the bio-energetic process, these words express exactly what happens.

Orgastic longing, which plays such an enormous role in animal life, now appears to express this "striving beyond one's own self," this "yearning" to escape from the narrow confines of one's own organism. Perhaps here lies the answer to the riddle of why the idea of dying is so often represented in the orgasm. In dying, too, the biological energy reaches beyond the confines of the physical sac in which it is imprisoned. Thus the irrational religious concept of "liberating death," of "salvation in the hereafter," finds its true basis. The function that in the naturally functioning organism is fulfilled by the orgasm in sexual superimposition appears in the armored organism as the nirvana principle or as the mystical idea of salvation. The religious, armored organism expresses it directly: it wants to "free the soul from the flesh." The "soul" represents the orgonotic excitation, the "flesh" the surrounding confining tissues. The concept of "sinful flesh" has nothing to do with these facts. It is a defense mechanism in the pornographic structure of the human animal.

In summarizing, we may emphasize the simplicity of functional laws of living nature as one of their main characteristics. Functions as widely separated as growth, locomotion, and genital excitation can be reduced to the common denominator of the relationship between mass-free orgone energy and orgone energy that has become matter. The variations of this functional identity (common functioning principle) result secondarily in terms of the location in which this relationship appears in the organism. The width of the sac and its position (at the fore or rear end) dictates whether the deflection of the orgonotic current is expressed as growth energy or as sexual energy. But, seen functionally, all subsequent functions of living matter originate in the primal contradiction between the material and the bio-energetic orgonome. On the basis of this contradiction in living matter, one is even tempted to trace the connections that form the transition to the "highest" contradictions between "materialistic" and "spiritualistic" philosophy. But such an undertaking transcends the competence of this investigation and must be left to further research.

We shall encounter again the function of orgonotic superimposition in the natural realms of biochemistry and astrophysics. For it is orgonotic superimposition that connects the living organism with nature surrounding it. Living matter arose from inorganic nature as a special variation and, in its superimposition, is functionally identical with it. From here the path leads to the orgonometric investigation of the functional principle of nature per se.

Chapter V

SUPERIMPOSITION IN GALACTIC SYSTEMS

We are turning now toward the macrocosmic phenomena of orgonotic superimposition. The bridge from the microcosmic and bio-energetic to the macrocosmic realm is contained in the well-established principle of the "orgonomic potential." This basic function is sufficient to explain the growth of microcosmic into macrocosmic orgonotic systems. The first superimposition of two orgone energy units necessarily disturbs the equilibrium of the evenness of distribution of cosmic energy through formation of a *first "stronger" energy system.* This first stronger system from now onward attracts other, weaker units and thus grows. There is basically no limit to the growth of an orgonotic system except by solidification or freezing of energy into inert mass. This same principle also holds for living orgonotic systems. Solidification of the bone system demonstrates clearly the limitation of infinite growth in metazoa. Similarly, it may be assumed that the formation of a solidifying core in a macrocosmic system must impede its further growth.

However obscure the detailed functions of such growth still may be, classical astrophysical research has already clearly though unknowingly demonstrated that the *creation of certain galactic systems is due to superimposition of two cosmic orgone energy streams. Most "spiral galaxies" show two or more arms that unite toward the "core" of the total system.*

The following photograph of a spiral nebula was taken at the Mount Wilson Observatory, March 10 and 11, 1910, with the 60-inch reflector telescope (exposure 7 hrs. 30 min.). The nebula is numbered G9—M 101, NGC 5457 (cf. fig. 26).

Fig. 26. Messier 101, spiral nebula
(Mount Wilson photograph)

At least four arms are clearly discernible, and possibly five or six arms constitute the total system. There cannot be any reasonable doubt as to the spiraling motion depicted in the photograph. It is a most impressive picture of COSMIC SUPERIMPOSITION of more than two cosmic orgone energy streams. At the center we see the nearly circular form of the future "core" where the merger of the various streams takes place. It is the growing initial disc-like core of the galactic system.

Various opinions have been voiced in astrophysical litera-ture as to whether the arms of the spiral nebulae indicate a dissipation or unification of the galactic systems. At least one astronomer, Harlow Shapley, astronomer at Harvard Uni-versity, expressed the belief that the spiral nebulae with their arms indicate the beginning stage of a growing galaxy.[1] The existence of the orgone energy compels us to support and to qualify this view. It makes many features of the total picture of the spiral nebula comprehensible:

1. The unmistakable expression of the spiraling motion.

2. The rotation of the total system.

3. The superimposition and merger of two or more cos-mic energy streams.

4. The beginning solidification of the denser core.

5. The birth of a gravitational center of the total struc-ture.

6. The orgone energy envelope of so many heavenly bodies, which rotates faster than the material core.

---

[1] "The possibility that the end products of spirals such as ours may be spheroidal galaxies appears to be worth considering. It is proposed only as a working hypothesis. On such a plan, the evolutionary tendency among the galaxies would be from the Magellanic type to the most open spiral . . . ; and thence through the other spiral forms . . . to the elliptical and spherical systems. Recently we have found that spiral arms appear more as condensa-tions in great star fields than as ejections from a central nucleus. . . . The direction of development usually assumed, from compact spheroidal to open spiral, implies the appearance of supergiant stars and star clusters late in the history of a galaxy—an unlikely procedure it seems to me." (*Galaxies*, Blakiston Co., 1943, pp. 216 ff.)

7. The differentiation into a hard "core" and a "periphery" with an energy "field" of the heavenly orgonotic system.

Of course, countless problems remain unsolved. However, as a workshop form for future detailed inquiry the orgonomic hypothesis seems most promising and deserves to be tested by observation and measurement.

As an appropriate model for our future workshop tasks, the following assumption regarding the stages in the development of fixed star systems seems necessary.

First phase: Moving streams of cosmic orgone energy, still unformed, structureless, with little or no effective differences in density potentials, the "irregular" galaxy (cf. fig. 27).

Second phase: Mutual approach of two or more such cosmic orgone energy streams, followed by superimposition and formation of a *spiral nebula* with two or more arms (cf. fig. 26).

Third phase: Merger and fusion in the spiraling center followed by concentration and *micro*superimposition with the effect of creation of matter and a progressively hardening core, or nucleus.

Fourth phase: Formation of a disc-shaped or spheroidal galaxy; progressive slow-down of total motion; disappearance of the arms of the spiral form, as best represented by Spiral Galaxy NGC 4565 and by NGC 891 in Andromeda (Mount Wilson Observatory photograph, cf. fig. 28).

Our own galactic system, as manifested in the "Milky Way," still shows clearly the spiral form with two arms.

Fifth phase: Formation of a globular cluster which consists of already clearly differentiated single stars, densest toward the center of the total cluster (cf. fig. 29).

Here is the natural limit of our survey. It is, however, essential to allow the same functions that govern the formation of galactic systems also to govern the formation of sin-

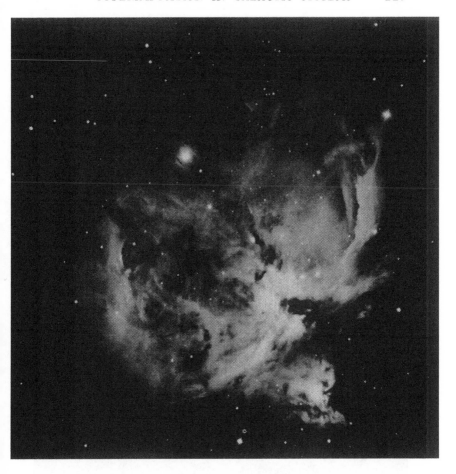

Fig. 27. "Irregular" galaxy (Mount Wilson photograph)

gle stars within the galaxy and of single planets around a fixed star. The ring of Saturn seems to demonstrate its origin from a disc-shaped concentration of orgone energy.

The basic form of the cosmic, galactic superimposition is the same as the basic form of organismic and microorgonotic superimposition; (cf. fig. 30, page 232).

Fig. 28. NGC 891, Andromeda, spiral nebula on edge
(Mount Wilson photograph)

Fig. 29. Messier 13, "Great Hercules Cluster"
(Mount Wilson photograph)

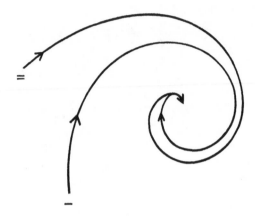

*Fig. 30. Cosmic superimposition of two orgone energy
streams*

The function of cosmic superimposition is most clearly visible in the following illustrations:
In the spiral form NGC 1042 (cf. fig. 31):

*Fig. 31. A drawing from Fig. 32 showing the direction of
flow of the two orgone energy streams*

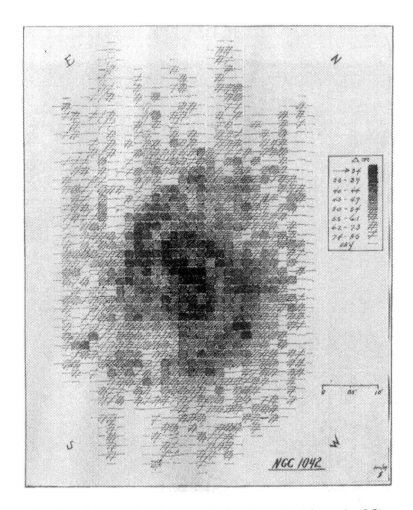

Fig. 32. A microdensitor analysis of a spiral form by Miss F. S. Patterson, working on a photograph made at Oak Ridge, according to Shapley's Galaxies

Here two cosmic orgone streams seem to be approaching each other from nearly exactly opposite regions of space.

In the spiral form NGC 1566 (cf. fig. 33) :

Fig. 33. NGC 1566, a southern spiral galaxy photographed with Harvard's southern reflector

Fig. 34. A drawing from fig. 33, showing the direction of
flow of the two orgone energy streams

Here the angle of approach is 180° less approximately
23°–25°.

In the spiral form G 10 (cf. fig. 35):

Fig. 35. Messier 81, spiral form G 10
(Mount Wilson photograph)

Here the approach is nearly exactly from opposite directions in a parallel manner (angle of approach, 180°).

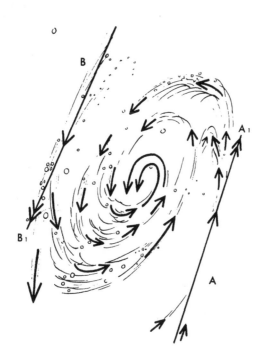

Fig. 36. A drawing from fig. 35 showing the direction of flow of the two orgone energy streams

These examples may, for the moment, suffice to demonstrate the high probability of the orgonomic work hypothesis regarding the creation of spiral nebular forms from superimposition of two or more cosmic orgone energy streams. Thus, not matter, particles or dust, but primordial orgone energy would constitute the original "stuff" from which galaxies are made. It is clear that this hypothesis tends to compete with the atomic theory, which places material particles

in the form of "cosmic dust" at the very root of cosmic creation. The orgonomic, energetic hypothesis requires that matter emerge from orgone energy through superimposition in
the microcosmic domain just as the whole galaxy emerges
through superimposition in the macrocosmic domain.

CHAPTER VI

THE RING OF THE AURORA BOREALIS,
R—76

Ever since the orgone energy was discovered in the atmosphere in 1940, it has become increasingly imperative to
find the concrete links connecting the orgone energy within
the living organism (bio-energy) and the (cosmic) orgone
energy outside the living organism. Long before the discovery of the atmospheric orgone energy, the aurora borealis
had been a major object of inquiry in orgonomic research.
From 1940 onward, however, this research acquired a systematic framework of thought and, thus, a clear direction.
The following facts and assumptions were the guiding lines:
    1. The unquestionable existence of a specific organismic
orgone energy quite logically forced the postulation of its
origin outside the living organism, somewhere in nature at
large. It was assumed that orgone energy develops from
matter, as in the bions. This was correct but far from complete. The existence of a mass-free orgone energy ocean that
fills the universe was unknown at that time. However, it was
already perfectly clear that the specific bio-energy inside the
organism must be derived from an identical energy outside
the organism. *How could bio-energy otherwise possibly have
gotten into the living system in the first place?*
    2. In 1939, it was established that orgone energy pos-
sessed the capacity of autogenous *lumination*. The specific

color of orgone energy in its natural state was already established as a *blue, blue-green,* or *blue-gray.*

3. *Pulsation,* as seen in living cells and organs, was a third important characteristic of orgone energy.

4. The functional thought technique (cf. *Ether, God and Devil*) was as yet far from its present state of development. However, attention was already directed toward the *pairing* of natural functions and the search for a *common functioning principle* (CFP).

5. It was clear that classical astrophysics had failed to comprehend the phenomenon of the aurora lumination. The word "ionization of the upper atmosphere" did not tell much, for where did the "particles" come from? And how was it possible that ions came down to the earth from the sun over a stretch of some ninety to a hundred million miles?

The first specific observations of the aurora borealis were made in Norway during 1937–1939 with little understanding. It was not until 1940, in Forest Hills, New York, that the observations gained systematic direction. The basic conclusion derived from many years of observation was the following: *the aurora borealis, or "northern lights," is the effect of orgonotic lumination at the outer fringes of the orgone energy envelope of the planet earth.*

Let us first describe the aurora lumination in its relation to basic orgone energy functions, which have become increasingly well known since the summer of 1940:

The color of the aurora borealis is generally an intense blue or blue-gray to blue-green. We know this color to be specific for most orgone energy phenomena. It can be easily observed microscopically in protozoa, cancer cells, bions of all kinds, and the frames of the red blood cells. The sky is blue. Deep inland lakes and the ocean are blue. So are thunderclouds. The lumination in charged vacuum tubes appears blue to the eye and blue on color film. The radiation from bions also shows blue on color film. In the metal-lined orgone room the lumination appears blue-gray at first, then increasingly blue up to intense violet. The soft glow of the firefly is blue-green. The "haze" on a clear sunny day in front of mountain ranges is blue. Blue are the sun spots, and also, at dusk, the flat valleys on the moon. A hurricane personally observed by the writer in 1944 was of a deep blue-black color. This enumeration may suffice at present.

The movement of the streamer type of aurora borealis is of a slow, undulatory, at times pulsatory and wavelike nature. Slow expansion and contraction as well as fast-moving protrusions as in the protoplasm of amoebae are characteristic of the aurora. This motion is of the same kind that can be obtained in highly orgone-charged argon tubes through excitation by a moving orgone energy field derived from the body or the hair. Furthermore, some aurora movements have a pushing or searching expression. This, of course, does not mean that these phenomena are life expressions. It only means that the same energy constituting the bio-energetic movements of pushing and searching is also present in the non-living realm of nature. It is necessary to mention this self-evident fact, since there are, especially among psychoanalysts with a bad conscience concerning orgonomy, strong tendencies to depreciate orgonomy with such remarks, among others, as "mystical" or "seeing blue lights and ghosts." Every observer is deeply impressed by the beauty and emotional impact of the flaming aurora. To experience it on silent nights is always exciting, quite different from the observation of a glowing radium dial (cf. fig. 38).

Color, movement and the rich emotional expression of the northern lights merge into one when they extend to nearly the total sky. The process is, on the average, this:

The aurora lumination usually starts at the northern horizon, at times directly at the horizon, at other times in a region some 20° to 30° above the horizon. In the latter case, very often a more or less regular arch, concave toward the earth's surface and sharply defined, separates the aurora from the northern part of the horizon. It is as yet entirely unknown what role is played by the North Pole in the typical emergence of the aurora in the north. Whether or not the huge iron deposits in the vicinity of the North Pole have anything to do with it would be hard to determine. This guess is based on the single observation of strong bluish lumination of the N-pole of a strong magnet held close to the metal wall of an orgone energy room.

The aurora lumination often remains low above the northern horizon without extending further. However, on many occasions the orgonotic excitation in the upper atmosphere drives the lumination into higher altitudes. If it lasts long enough, the lumination will tend, in the form of pulsating streamers, to reach out toward the region of the zenith of the observer, which is almost exactly 45° northern declination, according to the equatorial coordinate system at Orgonon, near Rangeley, Maine, (1800 feet above sea level).

We are now approaching the crucial point in the presentation of the aurora borealis: *"R—76."*

In describing the aurora borealis, the Encyclopaedia Britannica (1940) mentions this basic phenomenon with a few words only:

"Further north the direction of the dip needle approaches the observer's zenith and the corona effect is seen with rays spreading in all directions from this *central ring of light* and even reaching to the horizon." (Italics mine—W.R.)

This "central ring of light" in the region of the zenith constitutes our point of departure into deep secrets of the orgone energy in the universe. Careful observation of the "ring" of the aurora impels the observer to concentrate keenly on this phenomenon.

First: The rays of the aurora do not emanate from the ring toward the horizon. It is the other way around. When

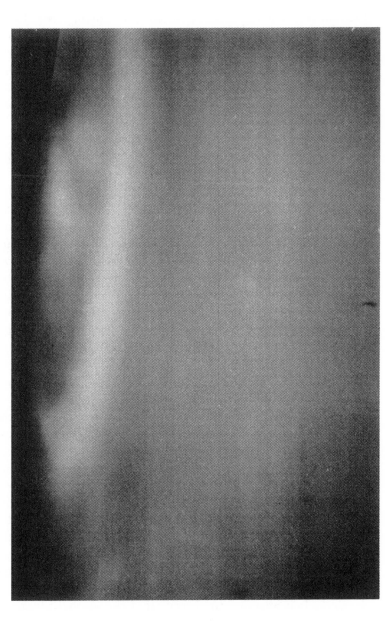

Fig. 38. Northern lights observed from Orgonon, Rangeley, Maine, September 5, 1950

the aurora, coming usually from the north, lasts long enough, it reaches with its pulsating digits toward the region of the zenith of the observer, which, at Orgonon, is almost exactly 45° northern declination. To the best of my knowledge, no attempt has heretofore been made either to define the exact location of the ring of the aurora or to comprehend this basic phenomenon.

Second: The "ring" (R—76) comes about only under certain well-defined circumstances. The aurora lumination must be strong and have a pulsatory, pushing quality. It must push upward in order to reach the region of the zenith. It must, furthermore, overshoot the zenith by a few degrees. Then, and only then will the southern part of the sky, too, begin to luminate and pulsate. This was observed for the first time on September 18, 1941, in Forest Hills, New York, and was later confirmed on many occasions (cf. table of aurora events, p. 246). The southern lumination usually develops first in the vicinity of the zenith. However, in spite of the fact that it later spreads farther south, the pushing, reaching-out movements are directed toward the zenith, and not away from it. Thus,

Third: The northern aurora lumination regularly induces lumination in the southern part, and

Fourth: Northern and southern lumination are directed against or toward each other. *They meet in a seemingly lawful spot several degrees to the south of the zenith of the observer at 45 degrees northern declination* at the longitude of Maine.

Fifth: The southern lumination, which is induced by the northern lumination at the zenith, and only under this condition, according to observations heretofore, does not seem to reach farther south than some 30 degrees. I have never seen the southern lumination reach down fully to the southern horizon.

Sixth: It is amazing to observe that the southern lumination disappears when the northern lumination recedes again from the zenith toward the northern part of the sky, and that it returns regularly when the northern lumination again is strong enough to reach beyond the zenith toward the south. This seems to justify the assumption that there are TWO *energy fields* in action, the *northern* and the *southern,* which excite each other into lumination.

Seventh: The "ring" itself does not always appear immediately as a ring. At times it does not form at all in a clear fashion. Instead, one can see that northern and southern aurora digits, in the process of their mutual approach, intertwine, winding around each other, receding slightly, intertwining again, merging and separating, merging again. This process, at times, takes the shape of a *spiral;* at other times a clear-cut circular *ring* is formed. It is, therefore, clear that the ring is formed by two streams of luminating energy. When the intertwining is most evident, the inside of the ring or the spiral becomes sharply delineated by the fact that it does not luminate. One has the impression that the center region of the ring escapes the excitation that otherwise induces the aurora phenomenon.

Eighth: Once the northern and the southern lumination have fully developed, the eastern and western sky are usually affected also and begin to luminate, until one sees a most impressive, moving, pulsating, cone-shaped, gothic-like *dome.* With the disappearance of the ring or the spiral at the zenith, the dome also slowly begins to vanish. The display of "ring" and "dome" lasted on a few occasions for from two to three hours, mostly shortly before and after midnight.

Before attempting to understand more of this amazing function, a table is presented summarizing those aurora displays that have been carefully studied. The number "R—76" denotes the appearance of a ring or a spiral. The + + + signs indicate by their number approximately the intensity of the phenomena. An analysis of the meaning of the number 76 will follow thereafter. R simply denotes "ring." Not all aurora displays between 1946 and 1950 were registered in the table, since not all of them have been sufficiently observed. All notations "R—76" marked with an asterisk rest upon measurements of the position of the ring by means of a celestial navigator star compass, of the variety used during the war by night fliers.

Observations of R—76 in Aurora Borealis (1946–1950)

No.	Date	Region of Origin	Form	R—76	Remarks
	1946	North	Dome		
1.	Aug. 30		Streamers	76-78°	
		South	Digits	+++	
	1946	North	Dome		
2.	Aug. 31		Streamers	+++	
		South	Digits		
	1946	North			
3.	Sept. 16		Streamers	+	
		South			
	1946	North			
4.	Sept. 17		Digits	++	
		South			
	1949		Arch		
5.	May 4	North	Digits	0	
	1949		Arch		
6.	May 5	North	Digits	0	
	1949				
7.	May 30 23.30— 23.45 h	*West- East!!*	Sharp Bands	++	*First* observation of W → E bands
	1949	From all		Tendency	
8.	June 5 23 h	directions	Curtain	only, incomplete	
	1949		Arch		
9.	Sept. 1 24 h	North	20-30° above horizon	0	
	1949		Arch		
10.	Sept. 25 21.45 h	North	Curtain	0	

No.	Date	Region of Origin	Form	R—76	Remarks
11.	1949 Oct. 14 21-22 h	North	Dome	++	
12.	1949 Oct. 15 18-19 h 22-24 h	*East-West* South also	Arch Dome	+++ ++++	Strong pulsation; *Arch bends northward;* R—76 persistent, brilliant; *superimposition perfect; spiraling*
13.	1949 Oct. 27 18.30 h	North East-West	Streamers No Pulsation	++	
14.	1950 May 27 23.30— 23.45 h	West- East	Parallel Bands	+++	
15.	1950 June 5 23.30 h	North	Curtain	+++	
16.	1950 Aug. 7 21.30 h	——	Streamers Pulsing	+++	
17.	1950 Sept. 5 22 h	North	Bands Curtain	0	
18.	1950 Sept. 8 3 h	North	Streamers	0	
19.	1950 Sept. 17 21.45 h	*East- West*	Narrow Band	Meeting of East with West at R—76 + no RING formation	*Eastern band bending northward toward galactic equatorial plane*

CHAPTER VII

THE MEANING OF R—76

The table shows that all registered ring phenomena have in common a position between 73° and 78° northern altitude or, in terms of the equatorial coordinate system, some 29° to 33° northern declination. This is a lawful occurrence of a natural function. Therefore, it must have some definable functional meaning. Let us now try to approach this lawful position of R—76 in an attempt to understand it functionally. WHAT CAUSES THE RING—76?

A few remarks are essential regarding the aurora display on May 30, 1949. It is of great significance that on May 27, 1950, the same type of display occurred exactly at the same time, between 23:30 and 23:45 h, with the ring, on both occasions, beginning to form at 23:30 h. On both occasions, furthermore, there was no northern or southern display. In both cases a thin, sharply drawn arch consisting of several narrow, luminating, straightly curved bands appeared in exactly a west–east direction. The ring was on both occasions formed by approach and contact of *two* band streams, one from the east, the other from the west, with the circular center in the usual position, R—76. In neither case did the center of the ring luminate. On both occasions, furthermore, the bands of lumination ran parallel, at times merging into one, at other times separating distinctly, but always held together, constituting one unitary west–east straightly arched band. The second display of this type, on May 27, 1950, lasted some ten minutes longer than the first on May 30, 1949. The color of the aurora was an intense blue. The west–east bands disappeared soon after the ring vanished.

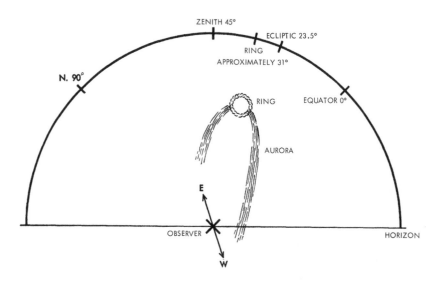

Fig. 39. Drawing of the west–east bands meeting at the ring of the aurora on May 30, 1949

THE AURORA DISPLAY ON OCTOBER 15, 1949, 18 to 24:30 HOURS

On October 15, 1949, at about 18 h, a complete, beautiful aurora borealis began to develop that deserves special description. It was preceded by an incomplete display on the evening of October 14. Both aurora phenomena were observed from the Orgone Energy Observatory at Orgonon, near Rangeley, Maine.

The aurora on October 14, 1949, developed in the north, about 21 h. It consisted mainly of streamers with little pulsation; however, there was a marked streaky structure and tendency to converge toward the zenith. The streamers did not, however, reach the zenith. A measurement of the position of the point of conversion in the region of the zenith, 45° N, gave approximately 76° northern altitude in the horizon coordinate system. Since, at Orgonon, 45° northern declination in the equatorial system corresponds to 90° northern altitude in the horizon system, the fictitious point of conversion of the northern streamers was located nearly exactly 31° northern declination: 45° plus 31° northern declination = 76° northern altidude.

The following evening at about 18 h, a much more general and also more intensive aurora began to develop. It started with a narrow band of lumination with sharp, even margins directed from *west to east*. Between 18 and 19 h, a region of circular broadening of the west–east band was observed. A rough measurement of the position of this broadening of the band gave approximately 78° northern altitude, i.e., 33° northern declination. In the course of the evening, this region of the luminating band changed its shape many times. Sometimes the western part of the band separated from the eastern part, joining again after a while. In the process of separation and rejoining, either a nearly complete ring formation, a circular disc, or a mutual torsionlike entanglement of the two bands developed. At any rate, this point was the most labile of the whole band, the rest being rather steady throughout the display.

There was no northern lumination in the beginning. Sometimes, one could see the west–east band bent toward the north as a whole. Later on, streamers set in from the south, 45° northern altitude, as well as from the north, toward a region exactly south of the zenith, which was located according to several repeated measurements at between 75° and 78° northern altitude, i.e., again between 30 and 33 degrees northern declination.

The display gained in intensity as well as extent. Two hours before midnight, a total "dome" all over the sky had developed. This dome was visible until after midnight. It was always centered, as in the tip of a cupola, at approximately 31° northern declination in a formation resembling a ring or a circular disc. Sometimes, when the streamers would recede for a short while, it was centered in a non-luminating meeting spot.

During the last two hours, extremely strong pulsations reaching out in broad pulsating bands toward the "ring" characterized the display, especially from the south, beginning at about 22° northern altitude, i.e., some 23° south of the equatorial plane.

The ring thus appears to be the area of contact between two streams of luminating orgone energy, northern and southern, or western and eastern. It is the product or result of *two* basic orgone functions, and we can, therefore, apply to it the orgonometric form of "creation," [1] which is

$$\mathbf{N} \mathrel{+\!\!<} \begin{array}{c} \mathbf{V_x} \\[4pt] \\[4pt] \mathbf{V_y} \end{array} \mathrel{>\!\!+} \mathbf{A}$$

N depicting the primordial cosmic orgone energy, V the principle of variation, x and y the two kinds of variations, and A the product of the superimposition of x and y, the ring of the aurora. Thus, A is known as the ring, but we must now try to comprehend the nature of x and y. We assume

---

[1] Cf. *Orgone Energy Bulletin,* October, 1950, "Orgonometric Equations, General Form," pp. 161–183.

they represent two separate orgone energy streams. Let us try to approach their characteristics concretely.

THE REALITY OF THE GALACTIC AND THE EQUATORIAL COORDINATE SYSTEMS OF THOUGHT

What is to follow now will astonish the reader; it has shaken the observer. Here, the objective validity of human thinking, if it follows logical sequences, will reveal itself with perfect clarity. Functional thinking will have to be acknowledged as one of the basic roots of nature in man, equal to his emotional and bio-energetic roots in the universe.

Orgonomy has deduced from observation of and reasoning about the type, direction and speed of the atmospheric orgone energy that there should exist an orgone energy envelope that not only surrounds the planet but also carries it along, just as water waves carry a rolling ball along in the direction of their progression. With this conclusion, an important view as to the nature of the MOVER of the planet has been found. We now know why the earth revolves and moves onward at all. It is being carried along by the EQUATORIAL ORGONE ENERGY STREAM. Let this stream be the y in our orgonometric equation. Let, furthermore, the ring of the aurora be A. Then two tasks result:

1. to establish the exact position of the ring with relation to the equatorial orgone energy stream;

2. to find the concrete meaning and quality of x, the pair of y in the orgonometric equation of "creation."

The measurements of the position of the aurora ring yielded, as an average, 76° northern altitude in the horizon coordinate system or, correspondingly, 31° northern declination in the equatorial coordinate system of classical astrophysics.

Since, in the equatorial system, the declination of the

equatorial plane is zero (0°), we obtain a function, the "ring," 31° north of the plane in which the equatorial orgone energy stream moves along.

If we now substitute 0 for the equatorial orgone energy stream and 31 for the ring, x, which is searched for, must have some relation to 0 and 31.

While pondering the nature of x, representing the second, unknown natural function, which pairs with y, representing the equatorial orgone energy stream, and results in the ring A, the *number 62 (degrees)* was found. The number *31 is the arithmetic resultant of 0 and 62,* if the latter represents vectors of two equal forces.

Beside the equatorial and the horizon system, astronomy uses the ecliptic system and the galactic system of coordinates in its calculations of the positions of heavenly bodies. The equatorial system of coordinates has as its X-axis the plane of the equator of the earth, extending into the celestial sphere. The horizon system has as its X-axis the (varying) horizon of the observer (at Orgonon, 45° northern declination). The ecliptic system is oriented regarding its X-axis according to the (apparent) path of the sun among the stars; it is inclined by 23.5° toward the equatorial plane.

In terms of actual movements this means: the earth rotates on its north–south axis in the equatorial plane, but it moves along in space "around the sun" on the ecliptic. Thus, the earth and all other planets do not move, as one would have to expect, with regard to the direction and momentum of their daily rotation in the equatorial plane, but they are being pulled northward with relation to their equatorial rotation to the amount of 23.5°. Naturally, the question arises: *what kind of force causes the deviation by 23.5 degrees to the north off the plane of daily equatorial rotation?*

According to well-known mechanical laws, such a force must exist, since otherwise the daily rotation and the forward motion (spinning) in space would have to occur in the same plane, namely in the equatorial plane.

This question had been in the mind of the writer for many years with no answer in sight. The position of the ring of the aurora provided the answer. Its position is about 7.5° north of the ecliptic; thus it could not be a function of the ecliptic itself, as was originally and tentatively assumed.

The number 62 finally solved the riddle in the following manner.

1. The Milky Way in astrophysics is viewed as constituting a plane along the galactic longitude, which is inclined by 62° toward the equatorial plane. The degree of the inclination of the galactic toward the equatorial plane thus provides the concrete number (62) that coincides with 31° north of the position of the ring of the aurora. The ring position (31°) thus appears the resultant middle between the *galactic* (Milky Way) plane, 62°, and the equatorial plane of the earth with the corresponding celestial equator, 0°. This is a numerical fact that unavoidably forces upon us major conclusions with regard to celestial mechanics. The following is a schematic presentation of what has just been said.

2. We have just performed a major operation of thought. WE HAVE SUPERIMPOSED TWO COORDINATE SYSTEMS, THE EQUATORIAL AND THE GALACTIC. This means: We have applied the function of superimposition, which is a concrete, real, observable natural function, to our thought operations proper. In astrophysics, the coordinate systems are merely used as imaginary frameworks of reference for astronomical measurements. No realities are assumed to be represented by these coordinate systems apart from the reality of the plane given by the Milky Way. Furthermore, in astronomy only one of the four coordinate systems is used practically to determine the position of a star in the heavens.

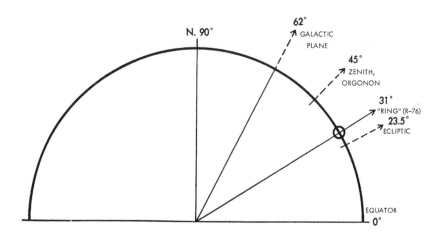

Fig. 40. Scheme depicting the angular relationship of "ring" (R—76) to galactic and equatorial plane

3. We have deduced a natural function, the aurora ring, with regard to its position at 31° northern declination, as a resultant of two other natural functions that are characterized by the numbers 0 and 62, the equatorial and the galactic planes respectively, as if they were realities of two cosmic forces. In brief, we have filled the two coordinate systems with energy doing concrete work in producing a concrete celestial phenomenon, the ring of the aurora. This happened in the course of a logical thought operation. The basic conclusion which follows is:

The second force, x in our equation, which pairs with the equatorial orgone energy stream, y, resulting in the ring of the aurora, A, is an orgone energy stream in the galactic plane, 62° off the equatorial stream.

4. We have filled all space of the Milky Way galaxy with

streaming orgone energy, and we have deduced the existence of two orgonotic streams within the planetary system, inclined toward each other at an angle of 62°. The path of the sun on the ecliptic, which deviates from the equatorial plane by 23.5°, thus appears as the resultant of the galactic and the equatorial orgone energy streams, with the latter constituting the slightly stronger force. This conclusion also requires that the plane of the galactic longitude be real, i.e., it is not only a circle running around the celestial sphere corresponding to the Milky Way, but it constitutes a plane that runs through the plane of motion of the planetary system. *The ecliptic, accordingly, would be the result of a pull exerted upon the planetary system by the galactic orgone energy stream,* making its course deviate from the equatorial plane by 23.5°.

5. A further inescapable conclusion is the following:

The equatorial orgone envelope provides the concrete physical mover of the planets. The planets rotate on their north–south axis and are carried along like rolling balls on progressing water waves, slower than the waves. The sun does not "attract" the planets. It rolls along in the same plane and in the same direction, carried along with the planets in the equatorial orgone energy stream. The path of the planets is neither a Copernican circle nor a Keplerian ellipse. It is of necessity open, and not closed, since there is a forward motion in space of sun and planets that never returns into its own path. Correlation of the classical astrophysical calculations, which use the circle and the ellipse, with the orgonomic "open path" of the course of the planets becomes now a major task of natural science. Since the path of the planets is necessarily a spinning wave, the coordination of classical and orgonomic astrophysical observations will have to deal with the integration of the Keplerian ellipse with the spinning wave.

At any rate, the PRIME MOVER of the heavenly bodies

has been postulated to be the cosmic orgone energy flow. We are now freed from the clumsy assumption that spheres revolve and move onward in an "empty space," i.e., with no actual physical forces responsible for the motion of the celestial bodies; with an ecliptic deviation of 23.5° from the direction of force of rotary spin for no apparent reasons; with sun and planets moving in the same direction in the same plane with no physical explanation for this lawful behavior.

Up to this point, we have deduced our conclusions from one single celestial function, the ring of the aurora, and its position in relation to the galactic and the equatorial planes. The rest was the result of thought operations, and not of direct observations. Though entirely consistent, the conclusions drawn with regard to celestial mechanics require more observational evidence. The solution to the riddle of the ring of the aurora was an important piece of evidence that led in an entirely unforeseen direction. New evidence was found hidden in a most unlikely spot.

FUNCTIONING OF HURRICANES

The following presentations are based on factual observations of hurricanes, compiled by Ivan Ray Tannehill, chief of the division of synoptic reports and forecasts, U.S. Weather Bureau, in Washington, D.C. (*Hurricanes*, Princeton University Press, 1945). The orgonomic presentation will restrict itself to those hurricane functions related to the main issue of this treatise—the existence of two concrete orgone energy streams, the equatorial and the galactic. Certain functions of the hurricane that have remained unexplained will become comprehensible from this point of view. No attempt at an interpretation of the hurricane "from the standpoint of orgonomy" will be made. The writer does not believe in and intensely dislikes "interpretations of unknown functions from this or that standpoint." He believes in the rule of not approaching natural functions with interpretations, but of "letting nature speak," i.e., letting the theoretical integration of various functions emerge from the natural processes themselves. The aurora ring had been observed for many years with no attempt at an interpretation, until it yielded its secret. The same is valid for the hurricane. It was not interpreted. It yielded its secret. It is hoped that this will emerge clearly in the course of the presentation of the subject. The author would like to add that he is not a professional meteorologist though he has studied weather functions in their relation to orgone energy functions in the atmosphere since 1940.

The emergence of the aurora ring from the function of superimposition of two cosmic orgone energy streams, corresponding to the equatorial plane and the galactic plane respectively, was known in its essential characteristics for years. The question was, how to prove this functional relationship. While this was being undertaken, basic aspects of the problem were elaborated.

To be in harmony with the theory, any new cosmic function would have to show clearly the function of superimposition, i.e., to be visibly a merger of two or more arms, as in the spiral galaxy. Its motion would have to be of a spinning nature, and, finally, it would have to agree with the assumed existence of two cosmic streams that cross each other at an approximate angle of 62°.

On August 22, 1949, a hurricane was in progress off Key West, Florida. It was photographed by Navy Photographer K. G. Riley with radarscope. This picture clearly demonstrated that the hurricane consisted of two arms that merged into the "eye," or "core" (cf. fig. 41). Careful analysis of the picture shows:

   1. *two* arms corresponding to *two* streams;

   2. approach from nearly opposite directions;

   3. curving inward of the two streams toward each other;

   4. intertwining and merger into one core, or eye: super-imposition.

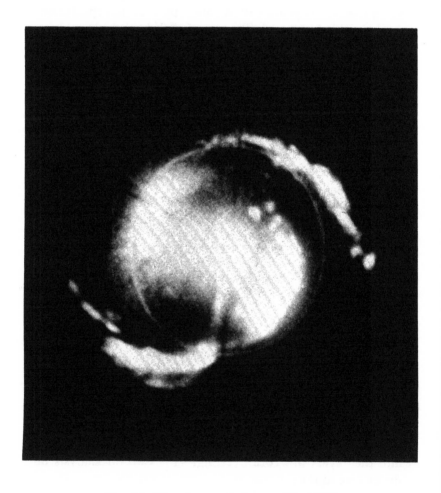

Fig. 41. Hurricane of August 22, 1949

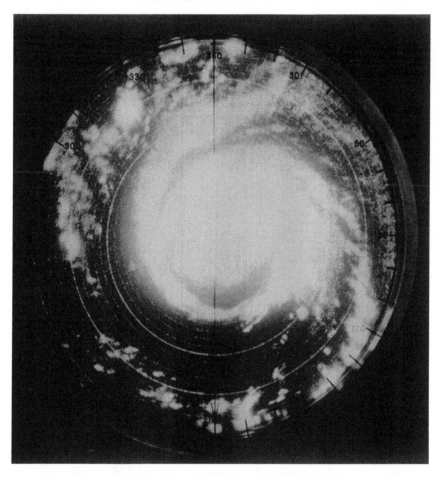

*Fig. 42. Hurricane of September 21, 1948, 11:31 a.m.
(U.S. Naval photograph, No. 706634)*

The U.S. Naval Air Station, Key West, Florida, kindly
provided from its files another radarscope picture of a hurri-

cane that developed on September 21, 1948, and was photographed at 11:31 A.M. off the Florida coast (cf. fig. 42).
This hurricane picture shows still more clearly the approach,
superimposition and merger of two streams, as in a galaxy,
forming a core with a counterclockwise spin.

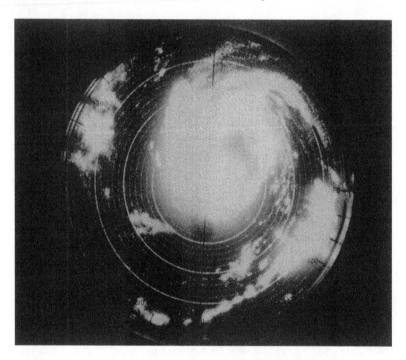

Fig. 43. Hurricane of September 21, 1948, 1:00 p.m.
(U.S. Naval photograph, No. 706635)

About two hours later the same day, at 1:00 P.M., another picture of the same hurricane had been taken (No.
706635). This second picture shows the two arms less
clearly, though still distinguishable; the "eye," or "core," on
the other hand, has grown. Direction of spin is the same, i.e.,
counterclockwise.

Thus, another proof of a natural function composed of two intertwining and superimposing streams was obtained. *The hurricane is a natural cataclysm, due to the superimposition and merger of two cosmic orgone energy streams.* The fact that the hurricane consists of two arms is not mentioned in Tannehill's book and seems to be unknown in the literature. If it is mentioned in the literature, I will be glad to correct my statement. Let us now return to other important characteristics of the hurricane.

DIRECTION OF SPIN AND FORWARD MOTION WITH REGARD TO NORTHERN AND SOUTHERN HEMISPHERE

It becomes perfectly clear from Tannehill's summary that there is a lawful behavior in the movements of hurricanes with regard to their origin in the northern and southern hemispheres. Tannehill writes under the heading "The Law of Storms," (p. 26, 1945):

On ships at sea passing through tropical cyclones, changes in direction and force of the wind are fully understood. A knowledge of the law of storms is an essential part of the education of ships' officers. To the landsman who experiences a tropical storm, the direction from which the wind blows, in relation to the position of the storm center, is sometimes puzzling. After the wind blows from one general direction for a considerable time, increasing in force, a calm succeeds, followed by a violent wind from nearly the opposite quarter. It simply means that the storm center has passed over the place. Nevertheless, it is frequently said that the "storm came back." When the wind blows from northeast toward the southwest, the conclusion is that the storm is coming from the northeast and moving toward the southwest. Then when the southwest wind succeeds the calm, the conclusion is that the storm has come back and is now moving from southwest to northeast. Such conclusions are altogether erroneous.

The wind is not "coming back," of course. The navigator had first passed one direction of the circular movement of the main body of the storm; then, after passage of the calm eye-center, he passes through the opposite direction of the flow. It has been established that hurricanes spin in the *northern* hemisphere in *counterclockwise* direction, and in the *southern* hemisphere in a *clockwise* direction. This lawful behavior has, to my knowledge, not been explained; it is obviously of extreme theoretical importance. Definite natural functions must be responsible for it.

The problem can be approached satisfactorily within the framework of the orgonomic postulation of two cosmic orgone energy streams that approach, meet, intertwine, superimpose, and merge. The sketch (fig. 44) will facilitate the understanding of the different directions of spinning north and south of the equator.

Let us assume that the two arms constituting the hurricane represent the equatorial and the galactic orgone energy streams. Then, their courses are inclined to each other at an angle of 62°. It does not matter in the present context whether we let the galactic stream flow from southwest toward northeast or from northeast toward southwest. With the equatorial orgone energy stream in both cases flowing from west to east, the counterclockwise spin in the northern and the clockwise spin in the southern hemisphere follow logically from the meeting of these two streams.

The sketch (cf. fig. 44) presents the first alternative, the flow of the galactic stream running from southwest toward northeast with an angle of inclination toward the equatorial west–east stream of 62°. In this case, the equatorial stream, before reaching the point of the crossing, will necessarily be drawn northward and the galactic stream westward. The two will approach and superimpose several degrees north of the equator and will merge in a counterclockwise spin. The momentum of the counterclockwise spin soon after the merger when it is strongest, will exert a force pressing westward. The total forward motion of the hurricane will at first necessarily be toward the west or northwest, i.e., it will be directed against the general direction of both the equatorial west → east and the galactic southwest → northeast directions. However, with progressive loss of momentum of torque, the hurricane will have to yield to the over-all west–east direction of the motion of the orgone envelope of the earth and will reverse its direction toward west–east. The turn, however, will be interfered with by the galactic northeastward stream. Thus, as a consequence, the hurricane will, in the northern hemisphere, sooner or later be forced into a more or less northeastern direction.

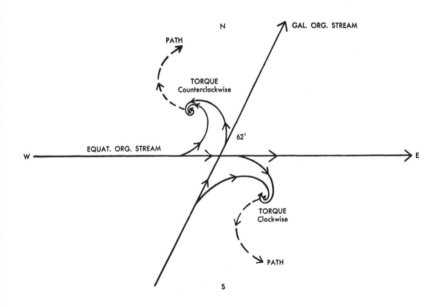

Fig. 44. Torque and path of hurricanes

This explanation of the original and the later direction of forward motion of the hurricane is in agreement with the factual observations as far as they have been registered and charted. The diagram (fig. 47), taken from Tannehill's presentation, will show that there is a lawful movement of most hurricanes in the northern hemisphere first westward or northwestward, in accordance with the counterclockwise momentum of spin, while, later, every hurricane that does not pass over toward the Gulf of Mexico into Texas (continuing in the direction of spin) will turn northeastward. An approximate measurement of the average degrees in the northeast turn yields the number 60° to 65° in near accordance with the angle of 62° between the equatorial and the galactic orgone energy flow.

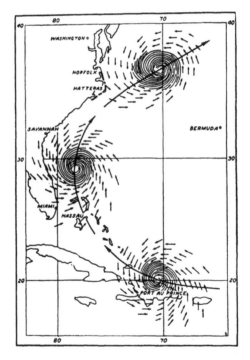

Fig. 45. Track and wind system of a tropical cyclone in the Northern Hemisphere (Tannehill's Hurricanes, p. 5)

This work hypothesis seems to be in disagreement with the concept that the solar system is no more than a tiny speck in the galaxy. We must leave this disagreement for further thorough scrutiny in a wider context.

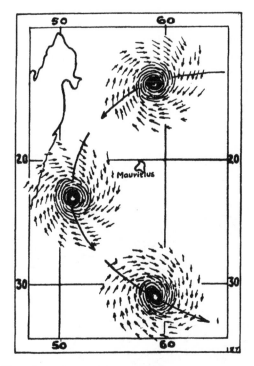

Fig. 46. Track and wind system of a tropical cyclone in the Southern Hemisphere (Tannehill's Hurricanes, p. 6)

CONTRADICTION BETWEEN DIRECTION OF SPIN AND DIRECTION OF GENERAL MOVEMENT

The counterclockwise spin in the northern hemisphere would, according to well-known mechanical laws, require a path of progression of the hurricane in, first, a westerly, then southwesterly, and finally a southern direction, since the path would curve in accordance with the curve of the spin. This is not the case. The general movement is opposed to the force and direction of the counterclockwise spin; it runs northward and finally northeastward. This requires a mechanical force acting against the direction of the spin, deviating from the path in exactly the opposite direction. Also, in the southern hemisphere this contradiction becomes evident from the charts of the paths of southern hurricanes. Here the clockwise spin should produce a curve from eastward toward southeast and finally south and westward. The oppo-

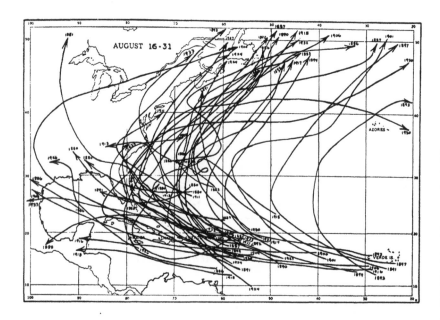

Fig. 47. Tracks of tropical cyclones of hurricane intensity, August 16 to 31, 1874–1933 (Tannehill's Hurricanes, *p. 66)*

site is factually the case. Southern hurricanes, as the sketch
(fig. 44) clearly demonstrates, run in the direction *opposed*
to the force of the spin.

The deviation in the northern hemisphere toward northeast can be explained by any one of three directions of
force opposed to the direction of the spin: by the galactic
orgone stream, which runs north–northeast 62° off the equator, by the ecliptic, which runs northeast 23.5° north, and by
the equatorial orgone envelope, which runs eastward in the
equatorial plane. In the southern hemisphere, the problem
remains open as to the southwestern, southern and final
southeastern directions. It seems safe to assume that the
turn eastward is due to the loss of momentum of spin with
final prevalence of the general direction of the rotation of
the globe.

THE SEASONAL OCCURRENCE OF HURRICANES

Another major problem arises in connection with the seasonal occurrence of hurricanes. It cannot be dealt with in this
context since the requirements for its discussion are much
too complicated. However, it is important to state the problem clearly.

The general course of rotation of the N–S-axis of the
earth's globe, the general direction of the postulated galactic
orgone energy stream in the plane of the Milky Way and the
resultant of the two, the ecliptic, remain the same. It is,
therefore, not comprehensible why hurricanes only arise at
certain seasonal periods, and not all year round. There must
be some good reason why the hurricane season in the northern hemisphere only lasts from about May to November,
with a steep peak of frequency in September, whereas the period from December to May is almost free of hurricanes in
the strict sense of "tropical storms" (Tannehill).

It can only be indicated that the problem just mentioned is of paramount importance in connection with celestial mechanics. Tannehill wrote upon special inquiry, and I quote with his permission from his answer of February 13, 1950:

> Poey's (the original compiler of the occurrence and frequency of hurricanes, 1856) list included some hurricanes between December and June. He listed five in January, seven in February, eleven in March, six in April and five in May. I omitted these from my list, because there is every reason to believe that these were not hurricanes, that is, storms of tropical origin. . . . In the last fifty years there have been a few storms of tropical origin in May and at least two have lingered into December, but there is no definite indication of storms of tropical origin in the Atlantic Gulf and Caribbean area during the months of January to April . . .

On the basis of carefully reasoned deductions we must agree with Tannehill's distinction between "tropical" and other storms. This distinction, with specially marked differentiation between the seasonal period May until November, and December until May, will prove to be of paramount importance for the understanding of certain functions in celestial mechanics. It will take time to formulate the whole set of problems that arise in a heuristically valid fashion. However, the natural function of superimposition of two or more orgone energy streams has opened up many avenues to the solution of basic riddles of cosmogony as well as celestial dynamics.

The following survey depicts the curve of frequency of tropical hurricanes from January to December, totaling 897 hurricanes in the northern hemisphere (Atlantic) that occurred from 1494 to 1944, compiled from Poey's and Tannehill's listings.

If we assume, as we must, that hurricanes are concrete manifestations of certain celestial dynamics, the sharp rise in frequency from about June to the peak in August and September should be carefully studied. For the time being, this exposition must suffice. The table was compiled by the author from material published in Tannehill's *Hurricanes*.

It is also significant that the average peak of hurricane activity falls in the month of September. September is the period when the ecliptic, the true path of celestial motion, approaches and crosses the equatorial plane from north to south at the autumnal equinox.

If our assumption is correct that the ecliptic represents the factual, astrophysical resultant of the equatorial and the galactic orgone energy streams, the meeting at the autumnal equinox is expressed concretely in the sharp autumnal rise in hurricane incidence. The next highest frequency falls appropriately in August, when the ecliptic approaches the autumnal equinox from the aphelion in June, the month lowest in the seasonal hurricane activity.

We must, accordingly, assume that:

1. Each planet possesses a disc-like orgone energy envelope that rotates faster than the globe.

2. All planets swing in a common galactic orgone stream, coordinated in time and plane of general motion.

3. Celestial functions such as sunspot cycles, aurora borealis, hurricanes, tides, major weather phenomena, etc., are immediate expressions of an interplay of two or more cosmic orgone energy streams.

The reader realizes that much can be expected from further elaboration of this base of astrophysical thought operation. From here, the path of inquiry leads directly into a reconstruction of the planetary movements in terms of *open*, spiraling, mutually approaching and receding pathways, and no longer in terms of closed elliptical curves.

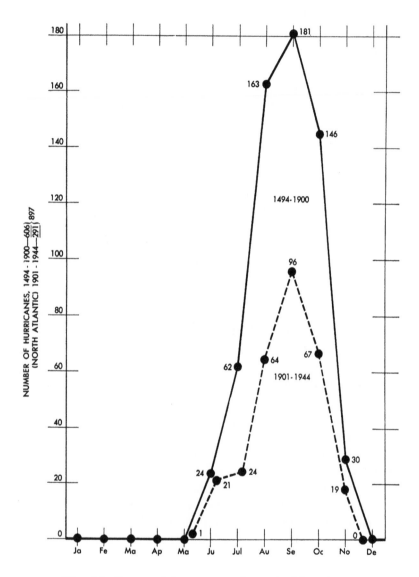

Fig. 48. Curve of frequency of tropical hurricanes from January to December

GRAVITATIONAL SUPERIMPOSITION*

Any attempt at an orgonomic theory of gravitation must proceed from functional and not mechanical principles. It will, first, have to abandon the absolute, eternal view of gravitation and replace it by the genetic view, according to which the basic natural laws themselves are being created and perish again as do all other natural functions. According to this view, *the gravitation of inert masses toward each other must have arisen with the creation of mass from mass-free primal cosmic energy.* It will, second, dispose of the mechanical gravitation of classical physics, not by mathematical abstractions, but by the closest possible observation of actual gravitational functions.

It is astonishing to witness the emergence of new insights from changes in theoretical principles. The functional principle of superimposition readily applies to the function of gravitation in the following manner:

The appearance of freely falling bodies misleads the observer into believing that an apple, for instance, falls vertically toward the center of the earth. This is true only with reference to the relationship: earth—apple. However, if we replace this immediate relationship of two variants by the CFP of both earth and apple, the orgone energy flow, we see that the apple actually does not fall vertically, and, furthermore, that it never reaches the center of the earth and never will, even if the substance of the earth would permit penetration toward the center. This is demonstrated by the behavior of the moon with respect to the earth. According to Newton, the moon appears to be falling like an apple toward the center of the earth. And the mathematical calculations corroborate this statement. Yet the moon never even reaches the surface of the earth. Thus, a gap exists between theory and appearance. The introduction of a *centrifugal* force that bal-

---

*The following is no more than a functional workshop arrangement.

ances the *centripetal* falling does not abolish this gap. It only complicates it by introducing a new unknown.

Let us follow the movements of both the moon and the earth with respect to their CFP, the flow of the cosmic orgone ocean. We are then in a position to integrate the appearance with the theory of mass gravitation.

The drawing (cf. fig. 49) demonstrates the factual interrelationship. Both moon and earth spin along in space, with their respective open (not closed) pathways mutually approaching and separating again. Therefore, *it is not the gravitational masses, but the* PATHWAYS *of the gravitational masses, which meet.*

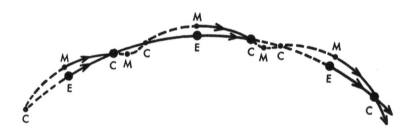

Fig. 49. "Moon (M) 'falls' toward center (C) of earth (E)"
Gravitational superimposition of paths of moon and earth

The moon does not "circle around the earth," since the lines of movements are open, spiraling curves. The moon does not reach the center of the earth. But it reaches actually a point in space where the center of the earth has been or will be sooner or later.

The cosmic orgone energy flow that carries both moon and earth along in the same direction, in the same plane, and in perfect coordination of their speeds, is the true agent of the gravitational free fall. It is with reference to the functional CFP of both earth and moon, the orgone energy stream, that otherwise contradictory statements about gravitation of heavenly bodies are proven to be true:

1. The moon actually falls toward the center of the earth. But it is also true that the actual material center of the earth is no longer present where, at that specific moment, the center of the moon passes through on its run "around" the earth.

2. It is apparently true that there is a "PULL" exerted by the earth upon falling bodies. But functionally, it is not due to a pull by the inert mass of the earth—which could never be demonstrated; *it is due to the primordial converging movement of two orgone energy streams.* This has been demonstrated in connection with the formation of galaxies. Gravitation as a function of converging streams of primordial energy is strongly suggested by the basic natural function of SUPERIMPOSITION of two orgonotic streams. Thus, it is again the CFP, the cosmic orgone energy stream, that accounts for the "gravitational pull."

3. It is also true that the actual falling of the moon toward the center of the earth is counteracted by an equal force acting in an opposite direction, resulting in an apparently circular motion around the earth's center. The moon never reaches the actual center of the earth, but it reaches the *virtual* center or, in other words, the point in space where the center of the earth had been a short while ago or will be a short while later.

It is fascinating to study these functions of nature as they are reflected in the searching mind of man, being true and untrue at the same time, depending on the point of view applied in the particular case.

To summarize:

The function of gravitation is real. It is, however, not the result of mass attraction but of the converging movements of two orgone energy streams. From these converging streams the "attracting" and "gravitational" masses once emerged and they are still carried along in the universe by the same streams in an integrated, unitary fashion as expressed in their common direction of movement, their common plane of motion, the mutual approach of their centers, and the mutually coordinated speed of their spinning motion (cf. fig. 50).

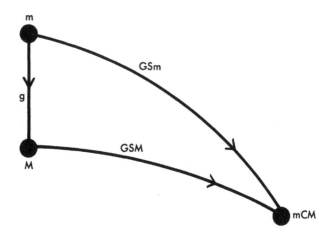

m, M . . . . two gravitating masses
g . . . . *apparent* gravitation
GSm, GSM . . . . *TRUE* gravitation through *SUPERIMPOSITION*
                        of ORGONE ENERGY STREAMS (PATHWAYS)
mCM . . . . m *apparently* at *CENTER* of M

Fig. 50. Gravitational superimposition

This seems to provide a solid framework for workshop activities for the future detailed elaboration and numerical definition of the planetary movements. The hypothesis of gravitation as a function of the superimposition of orgone energy streams is well deserving of close attention, paired with utmost caution and strict factual scrutiny of its realities.

CHAPTER VIII

THE ROOTING OF REASON IN NATURE

THE YEARNING FOR KNOWLEDGE

We have finished our surveying flight over the new territory abounding with knowledge yet to be harvested. We are turning homeward again, back into well-charted, familiar terrain. While we go over in our thoughts what we have seen unfolding beneath us, it may be well advised to ponder the greatest riddle of all: the ability of man to think and by mere thinking to know what nature is and how it functions. This ability is generally taken for granted. Yet it remains the greatest unsolved riddle so far. And on the solution of this riddle most probably depends the solution of the next-greatest riddle, the existence and perpetuation of the tremendous human misery for ages into ages. Men of knowledge do not feel called on to solve these riddles on order. They can only avoid as best they can the maze and entanglements of daily routine and *ad hoc* public opinion, and pursue their well-reasoned paths of search and thought.

There can be no doubt that rational thought and not political maneuvering, that hard, straightforward work on problems of existence and not mere voting, will open up the vastness of future human potentialities. It thus appears appropriate at the end of our flight to ask ourselves what place the human function of *knowing* may occupy in the scheme of natural events. We do not propose to enter into a complicated philosophical debate. We simply want to know what *knowing* itself does to man. So far, it seems to have done rather little to improve his lot. On the contrary, until now, the more he learned to know, the worse became the mass killing that has been one of the most horrible routines of daily life.

In pessimistic moods of hopelessness one is prone to ask what use there is in saving people from death by cancer if babies by the millions are being killed emotionally before and soon after birth in nearly every home all over the planet with the consent and help of their parents, their nurses, their doctors; when, furthermore, these emotionally deadened human babies later on, as grownups, carry any and every misdeed of cranks, politicians, dictators, emperors, and whatnot to evil power over men.

"SO WHAT?" From a biological and cosmic point of view it does not seem to matter at all; so goes one type of reasoning. Man has been maimed and killed by the billions over millennia. Whole species of living beings have arisen and perished. Civilizations have developed and vanished again. Religions have come and gone. Mighty empires that shook man's existence for centuries have crumbled, leaving no trace except a few ruins left over as witnesses of decay. So what? sounds in our searching minds again and again.

The cosmic orgone ocean, which has been surveyed in some detail in this book, pursues its eternal course whether we are aware of it or not, whether we understand the cancer scourge or not, whether the human race exists or not. It does not seem to matter. One understands well the mood of the retired and praying monk who lives only to return to God. Knowing about the cosmic orgone ocean, one has a better understanding of and feeling for the essentially ascetic nature of all major religious systems. Nothing matters . . .

Yet there lives and thrives in us a thirst for knowledge stronger than any philosophical thought, be it life-positive or life-negative. This burning urge to know can be felt like a stretching out of our senses beyond the material framework of our body, enabling us to understand what is rational in the metaphysical view of existence.

We yearn to know and to know better and with more certainty all the time, to pick up what those before us have learned, and to transmit it with our own small insights to the next and the following generations. We feel, in spite of all the "so whats" and "it does not matters," that we could not stop yearning for knowledge. We feel that we are tools of this yearning to know, as babies and puppies are tools of their plasmatic movements no matter whether or not there is sense and meaning in these movements. Seen from the bio-energetic standpoint, the human longing for knowledge obtains concrete meaning with regard to cosmic events.

The quest for knowledge expresses desperate attempts, at times, on the part of the orgone energy within the living organism to comprehend itself, to become conscious of itself. And in understanding its own ways and means of being, it learns to understand the cosmic orgone energy ocean that surrounds the surging and searching emotions.

Here we touch upon the greatest riddle of life, the function of SELF-PERCEPTION and SELF-AWARENESS.[1] This riddle is shrouded in awe; at times it results in frightened amazement, even complete confusion and disintegration of the searching ego, as in schizophrenia. All striving for perfection appears in this light as striving for the most complete integration of one's emotions and intellect; in other words, it is striving for the largest measure of bio-energy flow without blockings and deterring splits of self-perception. Therefore, the emotional merging in the genital embrace (pornography excluded), with unimpeded flowing of bio-energy, is most longed for and gratifying, as well as most beautiful in the aesthetic sense.

---

[1] Cf. "The Schizophrenic Split," *Character Analysis.*

In this light, and only in this, striving for perfecting knowledge has *cosmic* meaning. In penetrating to the greatest depth and the fullest extent of emotional integration of the self, we not only experience and feel, we also learn to understand, if only dimly, the meaning and functioning of the cosmic orgone ocean, of which we are a tiny part.

Since the "self" is only a bit of organized cosmic orgone energy, this full self-awareness is, seen from a deeper perspective, a step in the functional development of the cosmic orgone energy itself. Life energy has been defined as cosmic orgone energy, streaming within a membranous system. From this basic functioning all other and "higher" functions of the living system, including the intellect and the faculty of reasoning, emerge. Basically, the function of reasoning is not opposed or contradictory to the streaming of bio-energetic current. There is ample evidence in the biographies of great explorers, philosophers, and religious pioneers that their original reasoning grew out of experiencing their own life functions as cosmic events. And justly so.

Thus, in an ultimate sense, in self-awareness and in the striving for the perfection of knowledge and full integration of one's biofunctions, *the cosmic orgone energy becomes aware of itself.* In this becoming aware of itself, knowing about itself, growing into consciousness of itself, what is called "human destiny" is taken out of the realm of mysticism and metaphysics. It becomes a reality of cosmic dimensions, which merges understandably with all great philosophies and religions of and about man.

No great poet or writer, no great thinker or artist has ever escaped from this deep and ultimate awareness of being somehow and somewhere rooted in nature at large. And in true religion this was always felt, though never factually comprehended. Until the discovery of the cosmic orgone energy, this experience of one's own roots in nature was either mysticized in the form of transpersonal, spiritual images, or ascribed to an unknowable, forever closed realm beyond man's reach. This is what has always turned the quest for knowledge into mystical, irrational, metaphysical, superstitious beliefs. Thus, again "everybody is right in some way, only he does not know in what way he is correct." The discovery of the cosmic orgone ocean, its realities, and concrete physical manifestations such as the flowing of life energy in living organisms, puts an end to the compulsion of turning deeper searching into unreal, mystical experiences. The human animal will slowly get used to the fact that *he has discovered his God* and can now begin to learn the ways of "God" in a very practical manner. The human animal may well continue fighting his own full self-awareness for centuries to come; he may well continue to murder one way or another those who threaten his self-imposed blindness by orgonomic disclosures. As a mechanist or chemist, he will most probably defame this truly physical insight as a return to the phlogiston theory or to alchemy and, as a religious fanatic, he may well feel inclined to regard such a quest for extension of knowledge as a challenge to the greatness of the idea of an unknowable God, as criminal blasphemy. However this may be, the events cannot be reversed any longer. The discovery of the cosmic orgone ocean and its bio-energetic functioning is here to stay.

OBJECTIVE, FUNCTIONAL LOGIC AND MAN'S REASONING

The chain of events that unfolds during basic natural research demonstrates the *logic* of connections between various natural phenomena. The young research scientist experiences the unfolding of the logical chain of events as if there existed such a thing as "reason" in the universe. This is especially true when mathematical logic enters into the chain of sequences. It is most likely that the first ideas about an absolute "world spirit," no matter what you name it—in other words, the beginning of religious thought—emerged from man's capacity to observe and to reason about nature in such a fashion that consistent, objective logic emerged from this activity. We also have good reason to assume that at some time in the historic past the human animal was flabbergasted at this ability to follow logical chains of events that were beyond himself. What we are used to calling "objective natural science" is the summation of such chains of logical connections *beyond* ourselves.

Now this sounds like mysticism of the first order. The practical, technical business mind and the glibly brilliant intellectual are wont to sneer at such statements. However, they would fail completely in comprehending the fact that abstract mathematical reasoning is able to predict objective natural events. The deeply penetrating processes of basic scientific thought are foreign to them. So are the connections between deep intuition and crystal-clear intellectual elaborations of initially intuitive contacts with natural functions. So are, furthermore, such bio-energetic functions as the perfect care mothers give their offspring in the animal kingdom, the rational, logical activities of organs, most of the rational (objectively logical) processes in the growth of plants, the productions of a true musician or painter. To refer to these functions as the actions of an unconscious mind means nothing here. To identify "unconscious" with "irrational" is nonsense. The next question is inescapably: WHENCE STEMS THE UNCONSCIOUS MIND? And, if all functions below the conscious intellect are "irrational," how is it possible that life functioned well, long before the development of reason? There can be no doubt: *natural, objective functions are, basically, rational.*

The objective logic that leads from superimposition in the genital embrace to superimposition in the microcosmic (creation of matter) and in the macrocosmic realms (creation of the ring of the aurora, of hurricanes, and galaxies) stunned the discoverer and shook his emotions to their innermost depth. He has rejected the results of this logic for years and refused to believe that the conclusions to be drawn from them could possibly be true. For instance, he balked at admitting that true religion could be so very rational in spite of all its mystical distortions, that there could be such a thing as a rational core of all religious beliefs in an objective rational power governing the universe. But although he did not change his natural-scientific position and did not believe that a personified or absolute "spirit" governed the world, he found, more than ever, confirmation for the conviction that there exists and acts a physical power in the universe at the root of all being; a power, or whatever you may call it, that finally has become accessible to being handled, directed, measured, by man-made tools such as the thermometer, electroscope, telescope, Geiger counter, etc. While the discovery of cosmic orgone energy, the primordial creative force in the universe, was a triumph of enormous proportions, its importance hardly had the same emotional and intellectual impact that he experienced in the discovery of the workings of an *objective* functional logic in the natural functions beyond his personal being. In the midst of his emotional upheaval, he began to understand the absolute necessity of the idea of "God" among all peoples, whatever their race or whatever their kind of primitive awareness of this logic in nature may have been. It did not matter that the rational, logical chains of events in the universe had been so badly mysticized and personified; or that religious feelings and thought had been misused so often and so cruelly in the interest of secondary drives such as wars, exploitation of human helplessness, and misery, etc. *"God,"* at this point, *appeared to be the perfectly logical result of man's awareness of the existence of an objective functional logic in the universe.* Furthermore, it now appeared quite logical that man had again and again realized, in spite of all distortion and confusion, that somehow this same logic was functioning within himself. Otherwise, how could man possibly have become aware of the logic in nature outside of himself? How could he, furthermore, fail to become aware that he played a double role in the stream of nature: first, in realizing his ability to become actively aware of the logic in nature beyond his own self; and second, in spite of this ability, in being so badly and helplessly subjected to the powerful logic beyond himself, in birth and death, in growth and love, and, above all, in his insuperable drive toward the genital embrace. He must have felt right from the beginning that his genital drive made him "lose control" and reduced him to a bit of streaming, convulsing protoplasm. Here, the now well-known human orgasm anxiety may well have originated. It is no wonder, then, that most religions which tended toward monotheistic thought condemned the genital embrace through complete denial of all pleasure, as in the Buddhist religion, and by defamation of the genital embrace as "lust," as in the later Catholic religion. It is safe to assume that the impelling drive to overcome the basic natural function of the orgastic convulsion that rendered man helpless was later justified by the development of ugly, secondary, perverse, sadistic, cruel drives in man. The first struggles of the founders of many religions were quite obviously directed against these distortions of nature. Since no distinction between primary, natural genital drives and secondary, perverted, cruel, lascivious drives was yet possible, the most essential root of man in nature, his orgastic convulsion, fell prey to suppression, physiological blocking, and, finally, together with the secondary anti-social drives from which the primary drives were not distinguished, to severe condemnation.

In this manner, man "lost his paradise" (orgastic root in nature) and fell prey to "sin" (sexual perversion). He lost contact with one of his most crucial roots in nature and thus with nature itself, not only in the sensory and emotional but also in the intellectual realm. He could neither be in contact with nor understand nature, except in devious, mystical ways or by abstract reasoning. In higher mathematics a few human animals retained a bit of natural contact with logic in objective nature, and they stood out as particular and prominent minds separated from the rest of mankind, which had lost its sense of natural functions. Furthermore, life, God, genitality remained as if forever tabooed, inaccessible, unreachable, whether they were glorified into heaven or condemned into hell. The ambiguity of hell and heaven, God and devil, their mutual interdependence and exchangeability remained a basic characteristic of all moral theology. This sharp antithesis was reflected in many other dichotomies over the millennia, such as nature versus culture, love versus work, etc.

Let us not follow this line of sequences further. It has been dealt with on many occasions, in many different contexts of human pathology, sociology, ethnology, in early orgonomy, as well as in many other branches of human knowledge. The only additional piece of insight to be secured in this study is the basic identity between objective logic in nature, as it meets man's senses, and the power of reasoning itself within man. Expressed in terms of our orgonometric, functional language:

$$
\text{Natural processes} \left.\mathrel{\Large+}\!\!\!\!\mathrel{\Large<}\right\{ \begin{array}{l} \text{objective functional logic of orgone energy} \\[1em] \text{subjective functional, logical reasoning on} \\ \text{the basis of orgonotic self-perception} \end{array}
$$

To repeat: The discoverer of the primordial orgone energy, which functions within man (*bio-energy*) and outside man (*cosmic primordial energy*), found himself confronted with this functional identity of objective and subjective natural logic. He felt himself a tool of this logic, a very active and faithful tool. He followed it wherever it led him, with awe and a deep sense of responsibility as well as humility. The functional identity of biological and cosmic superimposition was the result of this symphony of outer and inner natural logic.

What basic function, then, has the discovery of the cosmic orgone energy in the flow of natural development?

It is not empty speculation to determine one's place in the stream of natural events. What is specifically meant here is not the fact that man as an animal grew out of the cosmic evolution. The question here is what the process of the discovery of the orgone energy flow inside and outside man entails for his place in and his handling of nature. Man is not only rooted in nature; he also perceives, tries to comprehend and use nature.

The overcoming of the mystification of nature will be a necessary consequence of the discovery of the primordial dynamics of nature. Is it then too much to say that *the discovery of cosmic orgone functions within the human animal may well represent a major evolutionary step forward in the direction of a functional unity of the flow of cosmic and intellectual developments, free of contradiction.*

Human history leaves little doubt that until this discovery man's intellectual activities functioned mainly in opposition to the cosmic energy. Partially, this opposition expressed itself in mystification and personification of the primordial mover and creator; in other respects, it expressed itself in the form of rigid, mechanistic interpretations of nature. This has been especially true in the last three centuries, during which the mechanistic, atomic, chemical view grew in opposition to the mystical distortion of nature. In *Ether, God and Devil,* an attempt was made to show that the primitive animistic view was closer to natural functioning than the mystical and the mechanistic. The mystical was overcome by the mechanistic; however, it never lost its hold on the minds of the majority of mankind. Both mysticism and mechanistics have failed as systems of thought. Mechanistics had to abdicate during the first half of this century, beginning with the discovery of nuclear radiation and Planck's demonstration of the quantum action at the basis of the universe. The animistic view, and not the mystical, was a forerunner of functional thinking, as expressed most clearly in Kepler's *vis animalis* that moves the heavens.

Orgonomy, at first without being aware of it, had picked up the thread that led in a hidden manner from the most primitive perception of nature by ancient man (animism) toward the establishment of the perfect functional identity between life energy (organismic orgone energy) and cosmic orgone energy. This identity of the two forms of existence, naturally, is a late development. Before man could ponder nature, he had to exist as an organized tiny part of cosmic orgone energy; and before he could exist, he had to develop out of a long series of predecessors. These predecessors, whether they pondered their origin or not, had to develop from very primitive plasmatic, orgonotic living beings that doubtless already possessed the ability to perceive and to react to the surrounding orgone energy ocean. This is merely a survey, to secure a firmer hold on our basic questions:

1. *Why was man the only animal species to develop an armor?*

2. *Was the armoring of the organism, which clearly is responsible for the mystification as well as mechanization of nature, a "mistake" of nature?*

The problem of why man was the only animal species to

develop an armor around his living core bothers the orgonomic educator and physician in his daily tasks. He has to
remove the armor in sick people and prevent the armoring
in children. In this difficult task he not only experiences
the terror that strikes when the armor is dissolved; he also
suffers from all kinds of dangerous attacks on his work and
very existence by people everywhere in his environment. If
nothing exists beyond the confines of natural processes, why
does the armoring of the human species exist at all, since it
contradicts nature in man at every single step and destroys
his natural, rich potentialities? This does not seem to make
sense. Why did nature make this "mistake"? Why only in
the human species? Why not also in the deer or in the chip-
munk? Why just in man? His "higher destiny" is, clearly,
not the answer. The armor has destroyed man's natural
decency and his faculties, and has thus precluded "higher"
developments. The twentieth century is witness to this fact.

Or is the process of armoring in man no mistake of nature
at all? Is it possible that the armor came about in some comprehensible, rational manner, notwithstanding its irrational
essence and consequences?

We know it is mostly socio-economic influences (family
structure, cultural ideas on nature versus culture, requirements of civilization, mystical religion, etc.) that reproduce
the armor in each generation of newborn infants. These infants will, as grownups, force their own children to armor,
unless the chain is broken somewhere, sometime. The present-day social and cultural reproduction of the armor does
not imply that when armoring first began, in the faraway
past of the development of man, it was also socio-economic
influences that set the armoring process into motion. It
seems rather the other way around. The process of armoring, most likely, was there first, and the socio-economic processes that today and throughout written history have reproduced armored man were only the first important results of

the biological aberration of man. The emergence of the mystical and mechanistic ways of life from the armoring of the human animal are too clearly expressed and too well studied to be overlooked or neglected any longer. With the breakdown of the armor, the outlook of the human being changes in such a basic and total manner, in the direction of contact and identification with his natural functioning, that there can be no longer any doubt of the relationship between armor and mysticism as well as mechanistics.

Still, the question of how the human animal, alone among the animal species, became armored remains with us, unsolved, overshadowing every theoretical and practical step in education, medicine, sociology, natural science, etc. No attempt is made here to solve this problem. It is too involved. The concrete facts that possibly could provide an answer are buried in a much too distant past; reconstruction of this past is no longer possible.

What follows now is more than empty speculation, since it is based on present-day and abundant clinical experience. It is less than a practicable theory, since it does not provide any better hold on the problem. However, it is interesting to follow a certain line of thought, to see where it leads and, finally, to reflect upon one's *ability to think* and to comprehend such things as the reality of two cosmic orgone energy streams that by superimposition produce hurricanes that spin counterclockwise north and clockwise south of the equator. Thus, our curiosity is well justified.

The development of orgonomy was guided throughout by the logical integration of natural functioning:

First: It was functional reasoning about the layering of human character structure that led to the deepest emotions confined in the armor.

Second: From the logical, functional peeling off of the armor layers resulted the discovery of the deeply hidden orgastic anxiety and the orgastic convulsion.

Third: It was reasoning about the *trans*personal and *trans*psychological nature of the orgasm function that disclosed its *bio-energetic* nature and the well-known fourbeat of the life formula : tension→charge→discharge→relaxation.

Fourth : It was functional reasoning again, more and more closely mirroring natural objective functions, that led from the *life formula* to the bions or energy vesicles and from there to the discovery of the radiation in bions, i.e., BIO-ENERGY.

Fifth : The same red thread of functional thinking led from the energy *within* living organisms to the same kind of energy *outside* in the atmosphere and from there further into the universe at large : COSMIC ORGONE ENERGY.

Sixth : Finally, it was again the orgasm function, abstracted into a generally valid natural principle, *superimposition,* that led to the understanding of the ring of the aurora and from there to the characteristic spin of multi-armed hurricanes and galactic nebulae.

The reader may well be aware of the fact that such a sequence could not possibly have been thought out arbitrarily. No human brain and no keen human fantasy could match this factual logic in the abundance of phenomena and interconnections, which yielded their secret to the natural observer who reasoned functionally.

This consistency of thought with the chain of the increasingly numerous natural functions that revealed themselves was no less amazing and at times even frightening to the observer who reasoned, than it must be to the reader of orgonomic literature covering a period of some thirty years. As the process of functional reasoning gradually unfolded, the observer not only worked out the method of this kind of functional reasoning; he also *experienced most vividly his own amazement at his own power of reasoning, which was in such perfect harmony with the natural events thus disclosed. The function of reasoning itself,* as part of natural functioning, *came to be a major object of consideration.* And here are some thoughts about the faculty of reasoning itself:

Before there was any life, there was the streaming of cosmic orgone energy. When climatic conditions were sufficiently developed on the planet, life began to appear, most likely in the form of primitive plasmatic flakes as reproduced in Experiment XX. From these flakes, single-cell organisms developed over the eons. Now cosmic orgone energy was flowing not only in the vast galactic spaces but also in tiny bits of membranous matter, caught within membranes and continuing to flow, still in a spiraling fashion, within these membranes, following a *closed* system of flow. We cannot assume that this tiny bit of streaming protoplasm already had developed the faculty of perceiving itself, although it already possessed the faculty of reacting to outer and inner stimuli. It was excitable, in agreement with the excitability of the orgone energy that flows outside the confines of membranes.

The confinement of a bit of cosmic orgone energy by and within membranes was the first clear differentiation of life from nonlife, of organismic from nonliving orgone energy. This much seems clear, even if it is as yet impossible to say much about the hows and whys of this genetic differentiation. Many years, unimaginable to human thinking, must have passed before this orgone energy, flowing within membranes in closed paths like the blood in higher animals, began to develop the faculty of perceiving its own flow, excitation, expansion in "pleasure," contraction in "anxiety."

We now have *three* streams of energy integrated with one another and emerging from one another: the *cosmic flow,* the *confined flow within membranes,* and the *first perception of the flowing itself,* i.e., ORGONOTIC SENSATION. A worm or snail might well represent the stage of development where sensation was added to objective plasma current. This orgonotic sensation is most clearly expressed in the drive to superimposition in the sexual process. Convulsion and discharge of surplus energy are already present. This phase must have lasted an immense period of time until it reached the stage of the higher animals. In a deer or an elephant, objective streaming of energy and sensation of streaming are still united. There is probably as yet no contradiction, no blocking, no wonderment; only pleasure, anxiety, and rage govern the bio-energetic scene.

Then man developed. At first, over long stretches of time, he was little more than an animal that had instinctual judgment, with the FIRST ORGONOTIC SENSE of orientation already in operation. There did not yet exist what we call *reasoned thinking*. This type of natural functioning must have slowly developed from the exact, sure contact between nature within and nature outside the orgonotic system. Whether or not the brain has anything to do with reasoned thinking, we do not know. The purposeful behavior of animals without a developed brain indicates that life does not require a fully developed brain to function properly. It is probable that reasoned thinking, in contradistinction to primitive, orgonotic reasoning as in all animals, somehow developed with stronger gyration of the brain. Since we generally assume that *functioning precedes and induces the structural development of organs,* and not the other way around, we must ask what kind of functioning forced the animal brain into a higher or more complicated form of existence. Whatever the answer to this riddle may be, man slowly began to reason *beyond* his strong orgonotic contact and harmony with nature, which heretofore had been sufficient to keep him alive and to develop him further, even into a reasoning being. We know nothing and can know nothing about those distant times when man began to think.

It is obviously wrong, however, to assume that thinking is a sharply distinguishing mark between animal and man. The transitions, to judge from natural processes in general, are always and everywhere slow, evolutionary, stretched over immense periods of time. *In the process of this development, man must have begun to reason about his own sensations of current and about his ability to perceive himself and to perceive at all.* To judge from the studies of the theories of knowledge, nothing can compare with man's amazement at his capacity to feel, to reason, to perceive himself, to think about himself and nature around him.

In thinking about his own being and functioning, man turned involuntarily against himself, not in a destructive fashion, but in a manner that may well have been the point of origin of his armoring, in the following way:

We know well from schizophrenic processes that an overstrained perception of self-perception necessarily induces a split in the unity of the organism. One part of the organism turns against the rest. The split may be slight and easily vanish again. Or it may be strong and persistent. In the process of this "depersonalization," man perceives his currents as an *object of attention* and not quite as his own. The sensation of bodily currents then appears, even if only in a passing manner, as alien, as coming somehow from beyond. Can we dare to see in this sharp experience of the self the first step toward mystical, transcendental thinking? We cannot tell exactly, but the thought deserves consideration.

There is much good reason to assume that in such experiences of the self *man somehow became frightened and for the first time in the history of his species began to armor against inner fright and amazement.* Just as in the well-known fable, the millipede could not move a leg and became paralyzed when he was asked and started thinking about which leg he puts first and which second, it is quite possible that the *turning of reasoning toward itself induced the first emotional blocking in man.* It is impossible to say what perpetuated this blocking of emotions and with it the loss of

organismic unity and "paradise." We know well the consequences of the blocking of emotional, involuntary activity: it
immobilizes the organism and disturbs the integration of all
biological functions. This may well have occurred when man
first turned his attention upon himself. From here on everything follows by its own inner logic of life-*negative* design
(cf. fig. 51).

The conclusion following from these thoughts is clear: *in
attempting to understand himself and the streaming of his
own energy, man interfered with it, and in doing so, began to
armor and thus to deviate from nature.* The first split into a
mystical alienation from himself, his core, and a mechanical
order of existence instead of the organic, involuntary, bio-
energetic self-regulation, followed with compulsive force. In
the brief sentence "Cogito, ergo sum" (I think, therefore I
am) the conclusion of one's personal existence follows from
the statement of the ability to think. The fright that still
overcomes man in our time when he thinks about himself; the
general reluctance to think at all; the whole function of re-
pression of emotional functions of the self; the powerful
force with which man resists knowledge about himself; the
fact that for millennia he investigated the stars but not
his own emotions; the panic that grips the witness of orgonomic investigations at the core of man's existence; the
fervent ardor with which every religion defends the unreach-
ability and unknowability of God, which clearly represents nature *within* man—all these and many other facts speak a clear
language regarding the terror that is connected with the
deep experience of the self. To stand aside, entirely logical
and dryly "intellectual," and observe your own inner functioning amounts to a splitting of the unitary system that only
very few seem to bear without deep upset. And the few
who, far from being frightened, enjoy submerging in their
innermost selves are the great artists, poets, scientists, and
philosophers who create from the depths of their free-

flowing contact with nature inside and outside themselves; in higher, abstract mathematics no less than in poetry or music. Are they now exceptions to the rule or the original rule itself? Is the majority of the human species the exception in the sense that it deviated from its unity with the natural orgone energy flow, whereas the few did not? It is perfectly clear that the basic answer to the misery of man depends on the answer to this question. For, if the majority represents what is natural and the few are the exceptions from the "normal," as so many want us to believe, then there is no hope of ever overcoming the split in the cultural setup, the wars emerging from this split, the splitting of character structures, the hate and universal murder. Then we would have to conclude that all the misery is a natural manifestation of the given, unalterable order of things.

If, on the other hand, *the majority is the exception from the natural,* and the few creators are in agreement with nature, then things would look better. It would become possible, by the most strenuous effort ever made in the history of man, to adjust the majority to the flow of natural processes. Then, if our exposition of the armoring is correct, man could return home to nature and what appears today as exceptional in a very few could become the rule for all.

It will be exactly those who suffered most from the deviation who will most strenuously object to the second possibility.

Here we encounter the possible effect of the discovery of cosmic orgone energy upon further human development in its fullest consequence. The discovery of bio-energy is here to stay. It will be opposed most severely by those who have lost contact with nature to the greatest extent. They will object. They will malign the discovery of life energy in the future as they have done for years in the past. They will defame the discoverer and the workers in the field of orgonomy. They will not shy away from any measure to kill

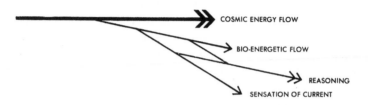

(I) MAN—ROOTED IN NATURE; CULTURE IN HARMONY WITH NATURE

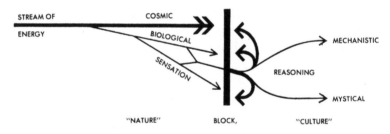

(II) MAN—DEVIATING FROM NATURE; "CULTURE" VERSUS "NATURE"

Fig. 51. Sketch depicting man's harmonious (I) and contradictory (II) rooting in nature

the discovery, no matter how devilish the means of killing may be. They will shy away only from *one* thing: from looking into microscopes or from doing any kind of observation that confirms the existence of an all-pervading cosmic energy and its variant, bio-energy.

In this process of fighting the discovery of cosmic orgone energy, a slow but most effective process of softening up the rigidities in the armored character structures will inevitably take place. The hardest, toughest, and cruelest character structure will be forced to make contact with the basic fact of the existence of a life energy, and thus, for the first time in the history of man, the rigidity in the human structure will begin to crack, to soften, to yield, to cry, to worry, to free life, even if at first in a hostile, murderous manner. The help of medical orgonomists will do its share in the softening-up process.

It is to be expected, furthermore, that as the public discussions of orgone energy functions spread over ever-widening areas of the globe, other human problems of existence will come into flux too. They will be subjected to a new type of scrutiny, and many gaps in understanding will be filled by what is already known about the basic cosmic force. The Catholic will have to revise his attitude toward the natural genitality of children and grownups; he will learn to distinguish pornography ("lust") from the natural embrace ("happiness," "body"). He already has begun to change his viewpoint with regard to the sexuality of children. Government officials will learn through sharp experiences in dangerous situations that man is far more than a *zoon politikon,* he is an animal with emotions that determine the course of history, irrational emotions to boot, which messed up the world in the twentieth century. One could even imagine that such rigid politicians as the Russian dictators would feel a "softening" toward human affairs creep into their frozen bodies. Religion will most probably revise its basic foundations regarding the sharp antithesis of man and nature, and will rediscover the real truth, which has been proclaimed with little factual knowledge or effect by most founders of religion throughout history. Work will enter the social scene as the toughest and most efficient combatant of political irrationalism. Man will learn to work for his life and love and children and friends, and not merely babble about the politics of the day, which are forced upon him by non-working parasites of society.

In this manner, the blocking of natural contact with the self and the surrounding world will slowly, possibly over several centuries, diminish, and finally, as the prevention of armoring in the newborn generations succeeds, will completely vanish from the surface of this earth.

This is no prophecy. Man, and not fate, is burdened with the full responsibility for the outcome of this process.

BIBLIOGRAPHY OF WORKS BY REICH ON ORGONE ENERGY

1. *Experimentelle Ergebnisse Über die Elektrische Funktion von Sexualität und Angst.* Copenhagen: Sexpol Verlag, 1937.
2. *Die Bione.* Copenhagen: Sexpol Verlag, 1938.
3. Communication to the French Académie des Sciences on Bion Experiment No. VI, January, 1937. Ch. IV.
4. *Bion Experiments on the Cancer Problem.* Rotterdam: Sexpol Verlag, 1939.
5. *The Discovery of the Orgone.* Vol. 1: *The Function of the Orgasm.* New York: Farrar, Straus and Giroux, 1973.
6. *The Discovery of the Orgone.* Vol. 2: *The Cancer Biopathy.* New York: Farrar, Straus and Giroux, 1973.
7. "Orgonotic Pulsation," *International Journal of Sex-Economy and Orgone Research,* 1944.
8. "The Schizophrenic Split," in *Character Analysis,* 3rd enlarged edition. New York: Farrar, Straus and Giroux, 1972.
9. "A Motor Force in Orgone Energy," *Orgone Energy Bulletin,* January, 1949.
10. "Further Characteristics of Vacor Lumination," *Orgone Energy Bulletin,* July, 1949.

11. "Public Responsibility in the Early Diagnosis of Cancer," *Orgone Energy Bulletin*, July, 1949.

12. "Cosmic Orgone Energy and 'Ether,' " *Orgone Energy Bulletin*, October, 1949.

13. *Ether, God and Devil*. New York: Farrar, Straus and Giroux, 1973.

14. "Orgonomic and Chemical Cancer Research. A Brief Comparison," *Orgone Energy Bulletin*, July, 1950.

15. "Orgonomic Literature Ordered from Russia," *Orgone Energy Bulletin*, July, 1950.

16. "On Scientific 'Control,' " *Orgone Energy Bulletin*, July, 1950.

17. "Orgonometric Equations: 1. General Form," *Orgone Energy Bulletin*, October, 1950.

18. "Meteorological Functions in Orgone-charged Vacuum Tubes," *Orgone Energy Bulletin*, October, 1950.

19. "The Orgonomic Anti-Nuclear Radiation Project (ORANUR)," *Orgone Energy Emergency Bulletin*, No. 1, December, 1950.

20. " 'Cancer Cells' in Experiment XX," *Orgone Energy Bulletin*, January, 1951.

21. "The Anti-Nuclear Radiation Effect of Cosmic Orgone Energy," *Orgone Energy Bulletin*, January, 1951.

22. "Complete Orgonometric Equations," *Orgone Energy Bulletin*, April, 1951.

23. "The Storm of November 25th and 26th, 1950," *Orgone Energy Bulletin*, April, 1951.

24. "The Leukemia Problem, I: Approach," *Orgone Energy Bulletin*, April, 1951.

25. *The Orgone Energy Accumulator: Its Scientific and Medical Use*. Rangeley, Maine: Orgone Institute Press, 1951.

26. "Armoring in a Newborn Infant," *Orgone Energy Bulletin*, July, 1951.

27. " 'Dowsing' as an Object of Orgonomic Research (*1946*)," *Orgone Energy Bulletin*, July, 1951.

28. "Three Experiments (*1939*)," *Orgone Energy Bulletin*, July, 1951.

29. "Wilhelm Reich on the Road to Biogenesis (*1935–1939*)," in *People in Trouble*. Rangeley, Maine: Orgone Institute Press, 1953.

30. *Cosmic Superimposition: Man's Orgonotic Roots in Nature*. New York: Farrar, Straus and Giroux, 1973.

31. "The Oranur Experiment: First Report (1947–1951)," *Orgone Energy Bulletin*, October, 1951.

32. *Contact With Space*. New York: Core Pilot Press, 1957.

GLOSSARY

ORGONE. Primordial cosmic energy; universally present and demonstrable visually, thermically, electroscopically and by means of Geiger-Müller counters. In the living organism: biological energy. Discovered by Reich between 1936 and 1940.

ORGONOMY. The natural science of the orgone energy; adj., orgonomic.

ORGONITY. The condition of containing orgone; the quantity of orgone contained.

ORGONOTIC. Qualities concerning the orgonity of a body or a condition.

ANORGONIA. The condition of lacking orgonity.

ORGONOMETRY. Quantitative orgone research.

ORGONOME.

Open: The form of a single spinning wave, viewed laterally.

Closed: An orgonometric figure obtained by bending an open orgonome in the middle and bringing together the ends. Specific form of the living organism. Prototype: egg shape. Characterized by the inequality of the two parts of the longitudinal axis, as distinguished from an ellipse.

303

INDEX